Engineering Practical Physics

Engineering Practical Physics

S.L. Kakani
M.Sc. (Physics), Ph.D.
Former Principal, Govt. SMM PG College, Bhilwara, Rajasthan
Former Convenor, Board of Studies and Research Board,
MDS University, Ajmer, Rajasthan
Director, MGIMT, Chittorgarh, Rajasthan

Shubhra Kakani
M.Sc.
Physics Department, Sukhadia University,
Udaipur, Rajasthan

CBS

CBS Publishers & Distributors Pvt. Ltd.
NEW DELHI • BANGALORE • PUNE • COCHIN • CHENNAI (INDIA)

ISBN : 81-239-1437-7

First Edition : 2007
Reprint : 2011

Published by Satish Kumar Jain and produced by V.K. Jain for
CBS Publishers & Distributors Pvt. Ltd.,
CBS Plaza, 4819/XI Prahlad Street, 24 Ansari Road, Daryaganj,
New Delhi - 110002, India. • Website: www.cbspd.com
e-mail: delhi@cbspd.com, cbspubs@vsnl.com, cbspubs@airtelmail.in
Ph.: 23289259, 23266861, 23266867 • Fax: 011-23243014

Branches:

- ***Bengaluru:*** Seema House, 2975, 17th Cross, K.R. Road, Bansankari 2nd Stage, Bangalore - 560070 Ph.: 26771678/79 Fax: 080-26771680 • e-mail: bangalore@cbspd.com
- ***Pune:*** Bhuruk Prestige, Sr. No. 52/12/2+1+3/2, Narhe, Haveli (Near Katraj-Dehu Road by Pass), Pune-411051 Ph.: +91-20-32404169 • Fax: 020-24464059 e-mail: pune@cbspd.com
- ***Kochi:*** 36/14, Kalluvilakam, Lissie Hospital Road, Cochin - 682018, Kerala • e-mail: cochin@cbspd.com Ph.: 0484-4059061-65 • Fax: 0484-4059065
- ***Chennai:*** 20, West Park Road, Shenoy Nagar, Chennai - 600030 e-mail: chennai@cbspd.com Ph.: 044-26260666-26202620 Fax: 044-45530020

Printed at :
India Binding House, Noida (UP)

Preface

This book is intended to meet the requirements of the practical curriculum for the students of B.E., B.Tech., B.Arch., A.M.I.E., Diploma and other engineering courses of all India and other Universities. Subject matter is developed in such a way as to be useful also to a wider readership.

The theory and the procedure associated with each experiment have been thoroughly described. The essential observation tables are given at the appropriate places. The precautions to be taken are also pointed out.

At the end of each experiment, some short question-answers are given for the better understanding of the subject and to help the students to prepare them for *viva voce*. This will induce self-confidence in the students and would serve to develop their insight.

Essential physical tables and log tables have been given at the end of the book.

We would like to thank Mr. H. S. Poplai, General Manager, CBS Publishers and his team for their untiring efforts in bringing out the book in time with excellent printing and nice get up.

Suggestions for the improvement of the book are most welcome.

S. L. KAKANI
SHUBHRA KAKANI

Preface

Contents

Introductory Concepts

General Instructions

Prior to performing an experiment, students are advised to read the following guidelines:

(*i*) *Study the experiment* very carefully, and think of steps you will take in carrying out the work.

(*ii*) *Check* that you have all the required *apparatus.*

(*iii*) *Keep a notebook* for a record of the experiment. In the notebook, write the *title* of the experiment and the date of your performance.

(*iv*) Then write the *working formula* with a clear explanation of the symbols.

(*v*) For an electrical experiment, *draw a circuit diagram* and make connections accordingly.

(*vi*) Do the work *cleanly* and *methodically*, step by step.

(*vii*) Record all the data in your notebook systematically in a *tabular form* as far as practicable. Provide each table with a suitable heading. Write at the top of each column of a table the quantity which is recorded in that column. Mention the *units* of the quantities.

(*viii*) Read the same quantity several times to *reduce random errors. Correct for systematic errors*, where known.

(*ix*) If a mistake is made in recording a value, pen through it and write the new value by its side. *Do not overwrite.*

(*x*) Draw graphs, where necessary.

(*xi*) *Calculate* the results, using *log tables* (or *electronic calculators*, if permitted). Put units after result, wherever necessary.

(*xii*) Compute the *percentage error*, and judge the number of significant figures to be kept in the result.

Take care of neat and clean recording of experimental data. Calculate the results desired and show essential steps of calculations wherever necessary.

Recording an Experiment in Practical Record Book

Normally, the entire stress is given to the actual record made by the student in the notebook that he uses in the class. Maintenance of a separate fair laboratory notebook is therefore not ordinarily required. In case, however, if a student is required to write up the experiment in a fair laboratory notebook after performing it, he needs some guidelines to do this. We suggest the following method of writing.

(*i*) **Title:** The title of the experiment should be written in block capital letters.

(*ii*) **Date:** The date on which the experiment was performed should be written after the title.

(*iii*) **Diagram:** A neat schematic diagram in pencil of the apparatus should be given on the left-hand page. In electrical experiments, a circuit diagram is required.

(*iv*) **Theory:** The theory of the experiment is to be outlined. Write the working formula and explain the symbols. Derivation of the formula is not necessary.

(*v*) **Apparatus or Experimental Set-up:** Give a complete list of apparatus and indicate their assembly, when required. A detailed description of an instrument or its action is not necessary.

(*vi*) **Procedure:** The steps followed in the work are to be mentioned.

(*vii*) **Experimental Results:** Record the data in the order they are taken. Make tables, wherever possible. Do not forget to write units.

(*viii*) **Calculations:** Calculations of the result by putting the measured values (in proper units) in the working formula are to be clearly shown on the left-hand page.

(*ix*) **Computation of Percentage Error:** A systematic calculation of the percentage error starting from the working formula should be given.

(*x*) **Precautions and Discussion:** Mention the precautions taken in the work. Discuss your results including comments on the measurements and the data.

Significant Figures

The measured value of a quantity is never *exact* owing to the errors involved. Reducing the errors, one gets an accurate value of the quantity, but it is not exact. In a measurement, the last digit is somewhat doubtful as it is always estimated. This doubtful digit is however significant since it gives an idea about the value of the quantity being measured. The figures that are reasonably reliable are known as *significant figures.* In the recording of experimental data, there is no reason for retaining a figure beyond the last significant figure which is somewhat uncertain.

Suppose that a voltmeter, graduated at intervals of 0.1 V, is used to measure a voltage in an electrical experiment. The reading is found to lie between 3.9 V and 4.0 V, and using eye-estimation it is recorded as 3.92 V. Any figure following the 2 here is meaningless since the 2 itself is estimated and is therefore somewhat uncertain. The reading 3.92 contains three significant figures. If the reading seems to coincide with the mark 4, we should record the value as 4.00 V. This way of writing emphasizes that we are capable of measuring up to three significant figures with the voltmeter. If we record the value simply as 4 V, it will imply that we measure up to one significant figure and hence the digit 4 is uncertain. But clearly there is no doubt about this digit in the measurement. Writing the value as 4.0 V is also incorrect as it sacrifices one significant figure. *The reading should be so recorded that it indicates the order of accuracy of the measurement.*

Drawing of Graph

If a change in the value of a quantity x causes a change in the value of another quantity y, we say that a *functional relationship* exists between x and y. If we use the functional relationship to determine y from a value of x, y is said to be the *dependent variable*, and x the *independent variable.* A visual display of the functional relationship between x and y is obtained by plotting them on a graph pater.

When the measured values of x and y are quite accurate and the functional relationship between them is known, it is easy to draw a graph by plotting the measured values. When the experimental data are not so accurate, one has to draw the 'best curve' that 'fits' the data.

A straight line represents the simplest functional relationship and is easy to draw. Even if the relationship between the quantities measured is not a straight line, mathematical manipulations may be made to obtain a straight line. Thus, if $y = ax^2$, or $y = A \log x$, where a and A are constants, we may plot y against x^2 or $\log x$, respectively, to obtain straight line graphs in the two cases. The best straight line that can be drawn from a given set of experimental data is determined by the *principle of least squares.*

Experiments

EXPERIMENT 1

Object

To determine the dispersive power of material for a prism for violet and yellow colours of mercury light with the help of spectrometer.

Formula

If A be the angle of the prism and δ_m be the minimum deviation of a ray of monochromatic light refracted through the prism in a principal section, then the refractive index of the material of the prism μ_λ for light of the given wavelength, is expressed by

$$\mu_\lambda = \frac{\sin \frac{1}{2}(A + \delta_m)}{\sin \frac{1}{2} A} \qquad (1)$$

If μ_1 and μ_2 are the refractive indices of the material of the prism for lights of two given colours and μ is the refractive index of the material for the mean colour between the two given colours, then the dispersive power of the material of the prism ω between the wavelength region considered is given by

$$\omega = \frac{\mu_1 \sim \mu_2}{\mu - 1} = \frac{\delta\mu}{\mu - 1} \qquad (2)$$

Equations (1) and (2) are used as the *working relations* for the determination of the dispersive power of the material of a prism.

Apparatus

Spectrometer, prism, sodium light, discharge tube, spirit level, mercury light source.

Procedure

(*i*) Adjust the spectrometer.

(*ii*) Determine the vernier constant of the circular scale and express it in minutes or seconds. Note that the vernier constant should not be in fraction of a degree.

(*iii*) Next proceed to find the *angle of the prism* as follows:

(*a*) Place the prism on the prism table with one of its refracting faces perpendicular to the line joining two of the levelling screws of the table and with its vertex at the centre of the table. Level it optically. Adjust the prism so that its vertex is directed towards the collimator and the axis of the collimator cuts the base of the prism nearly vertically.The parallel beam of light from the collimator will now fall simultaneously on both the refracting faces of the prism.

(*b*) See the image formed by reflection from one face of the prism with unaided eye and then set the telescope to receive the reflected beam (Fig. 1). Bring the junction of the cross-wire into coincidence with the centre of the image by moving the telescope slowly by its tangent screw. Note the readings of both the verniers and repeat this for three independent settings of the telescope.

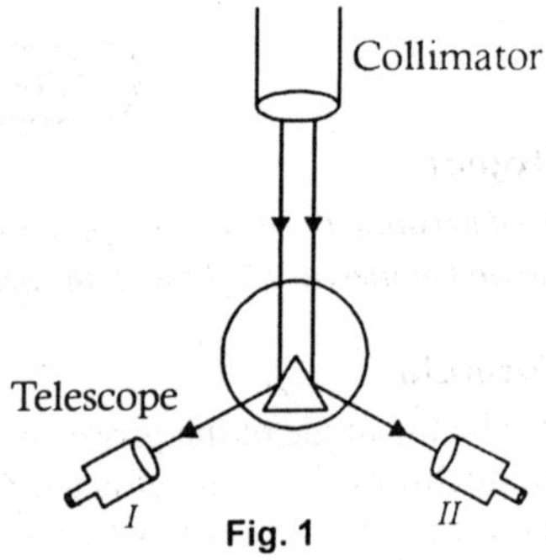

Fig. 1

(*c*) Repeat the operation (*b*) for the image formed by reflection from the other face of the prism.

(*d*) Find the angle between the two positions of the telescope. This gives twice the angle of the prism.

(*iv*) Next find the *angles of minimum deviation* for the two given specified colours and the mean colour. For this proceed as follows:

(*a*) Place the prism on the prism table with one of its face directed towards the collimator and the centre of the prism coinciding with that of the table. Set the prism at the position of minimum deviation for sodium light.

(*b*) Replace the sodium light by mercury light. Place the capillary portion of the discharge tube just opposite to the collimator slit so that the spectral lines as bright as possible. Adjust the prism at the position of minimum deviation for the yellow line of the spectra. Set the junction of the cross-wires at the centre of the line and note the readings of both the verniers. Take three independent sets of readings.

(*c*) Repeat the operation (*b*) for violet lines of mercury.

(d) Remove the prism without disturbing the table. Move the telescope in a straight line with the collimator to receive the direct light. Take the readings of both the verniers. The difference of this direct reading from the reading corresponding to the minimum deviation for a particular line gives the minimum deviation of that line.

Results

TABLE 1: VERNIER CONSTANT OF THE SPECTROMETER

.... Divisions Scale of Vernier (say, m)
= .... Division (say, n) of main scale

Value of 1 smallest main scale division (l_1) (cm)	*Value of 1 vernier division* $\left(l_2 = \frac{n}{m} l_1\right)$ (cm)	*Vernier constant,* $V.C. = (l_1 - l_2)$ (cm)

TABLE 4: DIRECT READING OF THE TELESCOPE

Vernier No.	*No. of observations*	*Main scale reading (M)*	*Vernier reading (V)*	*Total reading* $T = (M + V)$	*Mean reading*
1st	1	..	..	..	
	2	..	..	..	..(d_7)
	3	..	..	..	
2nd	1	..	..	..	
	2	..	..	..	..(d_8)
	3	..	..	..	

TABLE 5: DETERMINATION OF THE ANGLE OF MINIMUM DEVIATION

Colour of the line	*Angle of minimum deviation* (δ_m) *from*		*Mean* δ_m
	1st vernier	*2nd vernier*	
Mercury violet	$d_1 \sim d_7 (= e_1)$	$d_2 \sim d_8 (= e_2)$	$(e_1 + e_2)/2$
Mercury yellow	$d_3 \sim d_7 (= e_3)$	$d_4 \sim d_8 (= e_4)$	$(e_3 + e_4)/2$

TABLE 2: DETERMINATION OF THE ANGLE OF THE PRISM

Vernier No.	No. of observations	*Readings at the first position of the telescope*				*Readings at the second position of the telescope*				2A	Mean 2A	A
		Main scale (M)	*Vernier (V)*	*Total T = M + V*	*Mean T*	*Main scale (M)*	*Vernier (V)*	*Total T = M + V*	*Mean T*			
1st	1	..	..	..		..	..	..				
	2	..	..	..	(a_1)	..	..	..	(a_2)	$a_1 \sim a_2\ (=a)$		
	3	..	..	..		..	..	..				
2nd	1	..	..	..		..	..	..			$\frac{(a+b)}{2} = C$	$\frac{C}{2}$
	2	..	..	..	(b_1)	..	..		(b_2)	$b_1 \sim b_2\ (= b)$		
	3	..	..	..		..	..	..				

TABLE 3: READINGS OF THE TELESCOPE AT MINIMUM DEVIATION

Colour of the line	*Vernier No.*	*No. of observations*	*Main scale reading (M)*	*Vernier reading (V)*	*Total reading ($T = M + V$)*	*Mean reading*
Mercury Violet	1st	1	..	..	..	..(d_1)
		2	..	..	..	
		3	..	..	..	
	2nd	1	..	..	..	..(d_2)
		2	..	..	..	
		3	.	..	..	
Mercury Yellow	1st	1	..	..	..	..(d_3)
		2	..	..	..	
		3	..	..	..	
	2nd	1	..	..	..	..(d_4)
		2	..	..	..	
		3	..	..	..	

TABLE 6: DETERMINATION OF REFRACTIVE INDICES

Angle of the prism (A)	*Colour of the line*	*Angle of minimum deviation (δ_m)*	*Refractive indices (μ_λ)*
	Mercury Violet Mercury yellow	... ...	...(μ_1) ...(μ_2)

TABLE 7: DETERMINATION OF DISPERSIVE POWER OF THE MATERIAL OF THE PRISM

Refractive index for mercury Violet (μ_1) (from Table 6)	*Refractive index for mercury Yellow (μ_2) (from Table 6)*	*Refractive index for the mean Colour μ (from Table 6)*	*Dispersive power* $\omega = \frac{\delta\mu}{\mu - 1}$

Precautions

1. There should be no parallax between the cross-wires and the image of the slit.
2. The source should be properly aligned.
3. While measuring the angle of the prism, the prism should be placed on the table with its vertex coinciding with the centre of the table.
4. While rotating the telescope or the prism table from one position to the other sometimes it may so happen that the zero of the vernier crosses the zero of the circular scale. The student should note this carefully and, if so happens, he must find out (360°—difference of the two readings) to get the correct value of the angle turned through.
5. The mean colour between the two colours for which refractive indices of the material are μ_1 and μ_2 is that colour for which the refractive index of the material is approximately given by $\mu = (\mu_1 + \mu_2)/2$. Hence, if it is asked to find the dispersive power of the material of the prism between the wavelength region corresponding to two given colours for which the mean colour is not specified or easily identified, the refractive index for the mean colour can be found by the arithmatic mean of the refractive indices for the two given colours.

Questions with Answers

1. What is dispersion of light?

Ans. When a ray of white light passes through a prism, it is separated into the rays of its constitutent colours. This phenomenon is called dispersion.

2. Is there any other known device which produces dispersion of light?

Ans. Yes, grating also disperses white light.

3. Define dispersive power of a material.

Ans. It is the ability of a material to disperse various colours constituting the incident beam of light.

4. On what factors does dispersive power depend?

Ans. Dispersive power depends on the nature of the material and also on the wavelength for which it is determined.

5. Define angular dispersion.

Ans. For any two colours (wavelengths) angular dispersion is the difference between the deviations of those wavelengths (colours).

6. For the determination of the dispersive power, which is the best source of light?

Ans. Neon lamp, because it gives spectral lines in all the regions of the visible spectrum.

7. Does the deviation of a ray through the prism depend on the colour of light?

Ans. Yes, deviation of a ray through the prism is less for red then violet.

8. How does the deviation of a ray depend on the angle of the prism?

Ans. Greater the angle of the prism, more is the deviation of a ray.

9. What is the use of two verniers with the telescope?

Ans. In order to avoid error which may arise due to non-coincidence of the axis of rotation of the telescope with the centre of the circular scale.

10. Is there any limit for the value of the angle of prism?

Ans. Yes, when the refracting angle of the prism exceeds twice the critical angle for the material of the prism and air, no ray will emerge out of it.

11. Does the deviation through the prism depend on the length of the base of the prism?

Ans. Deviation is independent of the length of the base of the prism.

12. Why some prisms are made of larger base?

Ans. To increase their resolving power.

13. On what factors refractive index of the prism of the material depends?

Ans. μ depends on the pair of media and also on the wavelength (colour) of the light used.

14. How does the refractive index (μ) vary with wavelength?

Ans. Higher the wavelength, smaller the refractive index.

15. Ramsden's eyepiece is used in the construction of the talescope, how it is constructed?

Ans. Ramsden eyepiece consists of two plano-convex lenses each of focal length f separated by a distance equal to $2f/3$.

16. Why we use Ramsden's eyepiece with a spectrometer?

Ans. The cross-wire is outside the eyepiece and therefore, mechanical adjustment and measurements with this eyepiece are possible.

17. What are normal and abnormal dispersions?

Ans. In the normal dispersion the refractive index (μ) increases with the decrease of wavelength, whereas in the abnormal dispersion μ is higher for longer wavelengths.

EXPERIMENT 2

Object

To determine the wavelength (λ) of a monochromatic light with the help of Fresnel's biprism.

Theory

The Fresnel biprism is a prism which has one of its angles slightly less than two right angles and two equal small base angles. It acts like two very thin prisms placed base to base.

When rays from a slit, *S*, illuminated by a monochromatic light, such as sodium light, are made to be incident on the plane face of the biprism (*PQR*), the emergent rays from the two halves of the biprism appear to diverge from two coherent virtual sources, S_1 and S_2 (Fig. 2). If a screen (*AB*) is placed with its plane perpendicular to the plane containing the slit and the common base of the biprism, the emergent beams of light overlap on the screen producing alternate dark and bright fringes.

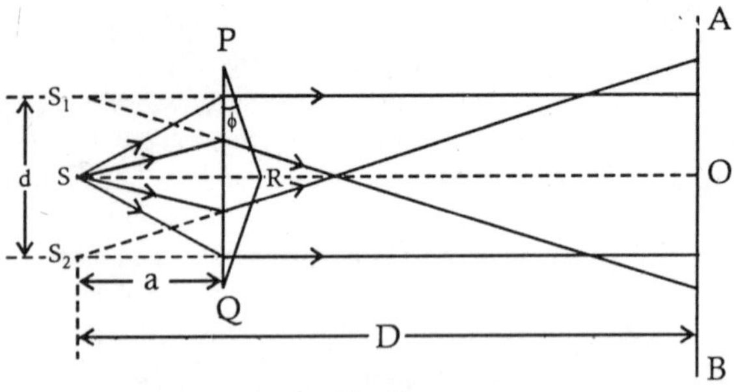

Fig. 2

If *d* is the distance between the two virtual sources S_1 and S_2, *D* is the distance between the slit and the screen, and λ is the wavelength of the monochromatic radiation, then the fringe width, *x*, i.e. the distance between two consecutive dark or bright fringes is given by

$$x = \frac{D}{d}\lambda \qquad (1)$$

In the actual experiment, the screen is the focal plane of an eye-piece, and the distance between the slit *S* and the focal plane of the eye-piece is obtained from the difference between the two vernier readings attached to the slit and the eye-piece stands. This distance gives an apparent distance D_1 between the slit and the screen, since there may be an index error between the slit and the eye-piece stands. If β represents this index error, then

$$D = D_1 + \beta \quad (2)$$

and the fringe width is given by

$$x = \frac{D_1 + \beta}{d}\lambda \quad (3)$$

Therefore, $\lambda = \dfrac{x.d.}{D_1 + \beta}$ (4)

Equation (4) is the working formula for the determination of λ; by measuring x, d, D_1, and β, the wavelength λ can be calculated from this equation.

Alternatively, the index error β can be eliminated by measuring the fringe widths x_1 and x_2 corresponding to two different apparent distances D_1 and D_2 between the slit and focal plane of the eye-piece. Then

$$x_1 = \frac{D_1 + \beta}{d}\lambda \text{ and } x_2 = \frac{D_2 + \beta}{d}\lambda \quad (5)$$

Hence, $\lambda = \dfrac{x_2 - x_1}{D_2 - D_1} d$ (6)

Equation (5) can also be used as the working formula to determine λ, and by measuring x_1, x_2, D_2 and d, λ be calculated from this equation.

When the quantities on the right hand sides of Eqs. (4) and (5) are measured in centimetres, λ is expressed in centimetres. In this experiment λ will be given in Angstroms, where 1 Å = 10^{-10} m.

To determine d, a convex lens having such a focal length that the distance between the slit and the focal plane of the eye-piece exceeds four times the focal length is interposed between the biprism and the eye-piece. The lens is adjusted so that for two of its positions the real images of the two virtual sources S_1 and S_2 are focused on the focal plane of the eye-piece. If d_1 and d_2 are the distances between the real images of S_1 and S_2 for two positions of the lens, then

$$d = \sqrt{d_1 d_2} \quad (7)$$

Apparatus

(*i*) A Fresnel biprism, (*ii*) a source of monochromatic light, (*iii*) a source of white light, (*iv*) a slit, (*v*) a micrometer eye-piece, (*vi*) a convex lens of suitable focal length, (*vii*) an optical bench with four stands, (*viii*) a plumb line, and (*ix*) an index rod (if required).

Procedure

The whole procedure may be divided into two parts: (*a*) *adjustments*, and (*b*) *measurements*.

Adjustments

(*i*) Mount the upright carrying the slit near the zero of the scale on the optical bench. Set up a plumb line near the slit and make the latter vertical by the tangent screw attached to it. Look along the axis of the bench from the other end and bring the slit in the middle of the bench by moving the slit-stand perpendicular to the bench. Make the slit very narrow with the help of the screw for adjusting the slit-width.

(*ii*) Mount the biprism on its stand with its plane face directed towards the slit and at right angles to the optical bench. Move the stand towards the slit until it is in contact with the slit-stand. Adjust the height of the biprism so that the centre of the biprism and that of the slit are at the same height.

(*iii*) Illuminate the slit with sodium light and adjust the position of the flame so that its brightest part is in front of the slit. Place a screen with a linear aperture between the slit and the source to cut off any stray light from reaching the eye. On looking at the slit through the biprism along the middle line of the optical bench, turn the transverse screw of the biprism till the common base of the biprism is found to move across the slit. This places the biprism in the middle of the bench. If there is an appreciable angle between the slit and the base of the biprism, adjust the tangent screw of the biprism till they are parallel.

(*iv*) Place the upright carrying the eye-piece on the optical bench and close to the biprism. Adjust the height of the eye-piece so that its axis may be at the same height as the centre of biprism. Focus the eye-piece on the cross-wires and move it perpendicular to the length of the bench until its axis passes through the centres of the biprism and the slit. Now indistinct fringes appear in the field of view. If, on looking through the eye-piece, you find neither a fringe system nor a bright patch of light in the field of view, turn the transverse screw of the biprism till it is found. Then make the fringes as distinct as possible by rotating the biprism in its own plane by the tangent screw.

(*v*) After obtaining well-defined fringes when the bases of slit-stand and the biprism-stand are in contact, proceed to check whether d_1 and d_2 will be observable. For this purpose, take the convex lens to be used for measuring d, and make a rough estimate of its focal length. Shift slowly the eye-piece away from the biprism and place it at a distance about 4 .5 times the focal length of the lens from the slit. While moving the eye-piece take care that the fringes never go out of sight. To keep the fringes in the field of view, adjust, the transverse screw of the biprism continuously.

Now, mount the lens on the optical bench with an upright between the biprism and the eye-piece. Adjust the height and the position of the lens so that its centre is on the line joining the centre of the slit, biprism, and the eye-piece. Move the lens along the bench to get real images of the two virtual sources S_1 and S_2 in the focal plane of the eye-piece for two positions of the lens.

(*vi*) Remove the lens from the bench, Slowly shift the upright carrying the eye-piece away from the biprism and place it near the end of the optical bench. While moving the eye-piece, keep the fringes always in the field of view by adjusting the transverse screw of the biprism. If the intensity decreases with increasing distance, adjust the slit-width, its position, and the intensity of the sodium flame. If the fringes are too broad, move the biprism slowly away from the slit till the fringe-width appears to lie within 1 to 2 mm. *Note that the biprism must not be moved to a distance greater than that corresponding to the position of the lens nearer to the biprism in the previous operation.*

(*vii*) Place and switch on a high-power electric lamp behind the sodium flame. Adjust its height so that the incandenscent filament is at the same height as the axis of the eye-piece. Coloured fringes with a white one at the centre will now be observed in the field of view. Bring the central fringe on the cross-wires by adjusting the transverse screw of the eye piece away from the biprism and bring it near the end of the optical bench. Keep the central fringe on the cross-wires by continuously adjusting the transverse screw of the biprism. Repeat these operations till the central fringe is on the cross-wires for all the positions of the eye-piece.

Measurement

(*i*) Remove the white light. Determine the vernier constant for the bench stands and the least count for the micrometer screw of the eye-piece. Fix the eye-piece at a distance of about 4.5 times the focal length of the lens from the slit. Place the stand carrying the lens on the optical bench between the biprism and the eye-piece, and make it co-axial with the latter. Now, on moving the lens-stand along the bench, magnified real images of the virtual sources S_1 and S_2 will be seen in the field of the eye-piece. Bring the images at equal distances from the cross-wires by rotating the lens about a vertical axis.This ensures that the images of both S_1 and S_2 will appear simultaneously in focus at each of the two positions of the lens. Measure the distance d_1 between two images by shifting the eye-piece perpendicular to the bench and setting the cross-wires on the same edge (left or right) of the image. Make this measurement four times, once shifting the eye-piece from left to right, next from right to left, again from left to right, and finally from right to left. The mean of these four values gives d_1. Readjust the lens in this position and repeat readings for two more independent setting of the lens. Take mean of these three values of d_1.

Next move the lens towards the eye-piece, keeping the latter fixed in its previous position, till sharp images of the virtual sources S_1 and S_2 are seen in the field of view. Determine the distance between the two images, d_2 in a manner similar to that followed in measuring d_1. From these values of d_1 and d_2, calculate d using Eq. (7).

(*ii*) Remove the upright carrying the lens from the bench. Move the eye-piece to a suitable distance from the biprism so that distinct fringes appear in the field of view. *Do not alter the position of the slit and the biprism*. Note the positions of the slit, biprism, and the eye-piece on the bench.

Turn the micrometer screw to shift the cross-wires of eye-piece beyond the last distinct fringe on one side (say, the left side). Then turn the screw slowly in the opposite direction to set the cross-wires in the middle of the 1st bright fringe on this side. Take the readings of the linear and circular scales of the micrometer. Next set the cross-wires on the middle of the 3rd or 4th bright fringe and take the linear and circular scale readings. The difference between these two readings gives the width for 2 or 3 fringes. In this way, go on measuring the width for 2 or 3 fringes (or more, if fringes are narrow) until the other end of the fringe system is reached. Again, starting from this side, and turning the micrometer screw in the opposite direction, measure the width for 3 or more fringes by similarly setting the cross-wires in the middle of the bright fringes. Take as many readings as possible for the width of three or more fringes and find the mean width of a single fringe.

When the index correction is to be avoided, repeat the above procedure to determine the fringe-width corresponding to another position of the eye-piece from the slit.

(*iii*) When λ is calculated from Eq. (4) index correction is required. To find index error, proceed as follows:

Remove the tube containing the cross-wires of the eye-piece from the socket. Also, remove the biprism from the bench. Take an index rod and put it parallel to the axis of the bench with one of its end touching the slit. Adjust the position of the eye-piece so that the other end of the rod is clearly seen through it. Take the readings of the indices attached to the slit and the eye-piece stands. The difference between these two readings give the apparent length of the rod. Measure the actual length of the rod by a meter scale, and find the difference between the actual and apparent lengths of the rod. This gives the index error.

TABLE 1: VERNIER CONSTANT FOR THE BENCH STANDS

..... divisions of vernier = ... division of main scale

Value of smallest main scale division (l_1) (cm)	*Value of 1 vernier division* $\left(l_2 = \frac{n}{m} l_1\right)$ (cm)	*Vernier constant* $V.C. = l_1 - l_2$ (cm)

TABLE 2: LEAST COUNT OF THE MICROMETER SCREW OF THE EYE-PIECE

Pitch of the screw p (cm)	*No. of division n on the circular scale*	*Least count* $= \frac{p}{n}$ (cm)

TABLE 5: INDEX ERROR BETWEEN THE SLIT AND THE EYE-PIECE

Posliton of the slit a_1 (cm)	*Position of the eye-piece when the end of the rod is in clear focus* a_2 (cm)	*Apparent length of the rod* $l_1 (= a_1 \sim a_2)$ (cm)	*Actual length of the rod* l_2 (cm)	*Index errour* $\beta = l_2 - l_1$ (cm)
...+...= ...	...+...= ...	...	...	...

TABLE 6: DETERMINATION OF THE WAVELENGTH λ

[For illustration, the table corresponds to determination of λ from Eq. (4)]

Fring width x (cm) [*From Table 4*]	*Distance* d (cm) [*From Table 3*]	*Apparent distance* D_l (cm) [*From Table 4*]	*Index error* β (cm) [*From table 5*]	$D_l + \beta$ (cm)	$\lambda = \dfrac{xd}{D_1 + \beta}$
...	...	...	...	...	...cm =...A

TABLE 3: DETERMINATION OF d

Position of the slit on the bench = ...+... = ...cm
Position of the biprism on the bench = ...+... = ...cm

Approximate focal length of the lens = ...cm
Position of the eye-piece on the bench = ...+... = ...cm

No. of obser-vations	*Position of the lens on the bench* (cm)	*Direction of eye-piece movement*	*Readings* (mm) *of the micrometer screw for*								*Distance* (mm) *bet-ween the images* $r_1 - r_2$	*Mean* d_1 & d_2 (mm)	$d = \sqrt{d_1 d_2}$ (mm)
			Left image				*Right Image*						
			Linear scale	*Cir. scale*	*Total* r_1	*Mean* r_1	*Linear scale*	*Cir. scale*	*Total* r_2	*Mean* r_2			
1	...	$l \to r$ $r \to l$ $l \to r$ $r \to l$	... ...	... ...	... ...	...	... ...	... ...	... ...	...	... = d_1 ...		
2	...	$l \to r$ $r \to l$ $l \to r$ $r \to l$								... = d_1		... = d_1	
1		$l \to r$ $l \to r$ $r \to l$ $l \to r$	... ...	... ...	... ...	...	... ...	... ...	... ...		... = d_2		
2	...	$r \to l$ $l \to r$ $r \to l$ $l \to r$								... = d_2		... = d_2	

TABLE 4: MEASUREMENT OF FRINGE-WIDTH

(*a*) Position of the slit on the bench = ...+... = ...cm
(*b*) Position of the biprism on the bench = ...+... = ...cm
(*c*) Position ofthe eye-piece on the bench = ...+... = ...cm

Apparent Distance between the slit and the eye piece (cm)	*Direction of the eye-piece movement*	*Eye-piece set at the fringes*	*Readings* (mm) *of micrometer screw*			*Width for three fringes* R (mm)	*Mean* R (mm)	*Mean fringe width* x $\alpha = R/3$ (mm)
			Linear scale	*Circular scale*	*Total*			
(*c*) ~ (*a*) =... (= D_1)	$L \to R$	1	...	...	...	...		
		4	...	...	...			
		5	...	...	...	...		
		8	...	...	...			
		...	...	...	...	...		
		...	...	...	...			
	$R \to L$	13	...	...	...	...		
		10	...	...	...			
		9	...	...	...	...		
		6	...	...	...	...		
		5	...	...	...			
					...			
		2	...	...	...			

1. In the measurement of the distance (d) between the two virtual sources S_1 and S_2, the distance between the slit and the screen should be nearly equal to 4.5 times the focal length of the convex lens so that d_1 and d_2 do not differ largely. This will reduce the error in measuring d.
2. While using the micrometer screw, care should be taken to avoid backlash error. Since there is often a considerable play between the micrometer screw and the nut in an old instrument, it is better to find the distance from the initial and final reading of the screw and counting the number of camplete turns, without depending on the linear scale.
3. While measuring d_1 and d_2, the images may be distorted due to spherical abberation. To avoid this distortion, a stop with a passage of light through the central portion of the lens may be used.

Questions with Answers

1. What is interference of light?

Ans. This is a phenomenon in which equidistant bright and dark fringes are produced by the superposition of light waves from two coherent sources.

2. What do you understand by coherent sources?

Ans. When two sources emit waves which have constant phase relatiohship between them at the origin, the sources are said to be coherent.

3. In order to produce sustained interference, why it is necessary that the two sources must be coherent?

Ans. Light waves from two sources traverse a distance before they superpose on a screen. Due to the path difference in the two rays from two sources to the point of interference, a phase difference is introduced between them. If the total phase difference between the two rays superimposing at the point be an even or an odd multiple of π , then the point will have a maximum or a minimum intensity. The phase difference due to path difference remains constant at the point. Now, if the initial phase difference between the two rays be constant, then and only then the point will have a sustained maximum or minimum intensity. If the phase difference between the two rays at their origin changes with time, i.e. if the two sources are not coherent, the point will have alternatively maximum and minimum intensity and as a result, no sustained interference will be produced.

4. How two coherent sources are realised in the Fresnel's biprism experiment?

Ans. The two virtual sources produced by refraction of light from a single source of light through the two halves of the biprism, serve as two coherent sources of light in this experiment.

5. What will happen if white light is used instead of monochromatic light in Fresnel's biprism experiment?

Ans. We know that white light is a combination of seven colours of which the violet has the least wavelength and the red has the greatest wavelength. Path difference between the two interfering rays at the centre is zero for all wavelengths. Clearly, the central fringe will be bright. The spacing between the two consecutive bright or dark fringes decreases with the decrease of wavelength $\left(\because x = \frac{D}{d}\lambda\right)$. Obviously, the bright fringe nearer to the central fringe will have strong violet tinge, followed by other bright fringes having a strong tinge of colours in the spectral order. After some fringes the path difference between the two interfering light waves becomes so large that the conditions for constructive and destructive interference may be satisfied at a given point simultaneously for a number of wavelengths and as a result one finds that the dark fringes of some wavelengths are completely marked by the bright fringes of other wavelengths. Moreover, at some points, so many colours may superpose over each other that the resultant illumination essentially appears white.

6. What will be the effect if we use broaden slit in place of narrow slit?

Ans. When we gradually broaden the slit, the image system first becomes blurred, because a broad source will serve as many narrow sources; each of these narrow sources will produce two coherent virtual sources and the fringe system due to one pair of virtual sources will not be formed exactly in the same position as that due to other pairs causing a blurred fringe system.

However, when slit is sufficiently broadened, the dark fringes due to one pair of virtual sources will be completely masked by the bright fringes due to another pair of virtual sources will be completely masked by the bright fringes due to another pair. As a result, one finds that no fringe system will be observed.

7. Is it possible to locate the central fringe?

Ans. The bright fringes for the sodium light have the same colour and it is difficult for one to locate the central fringe. However, for white light the central fringe is white and it may be located.

8. Can you determine the biprism angle ϕ ? What is the order of ϕ?

Ans. ϕ is the of order of about 30' when ϕ is increased, d increases and fringes become narrower. Using the relation $d = 2a(\mu - 1)\phi$, one can determine ϕ.

9. What are the basic differences between the biprism fringes and Newton's rings?

Ans. Biprism fringes are formed by division of wavefront and are straight, whereas Newton's rings are formed by division of amplitude and are circular.

10. Would you prefer to have d_1 and d_2 nearly equal?

Ans. Yes, using d_1 and d_2 nearly equal will reduce the error in the measurement of d.

EXPERIMENT 3

Object

To determine wavelength of sodium light by Newton's ring method.

Apparatus

Sodium lamp, short focal length lens, travelling microscope, Newton's rings apparatus having a plano-convex lens of focal length about 100 cm, two glass plates; and a spherometer.

The experimental arrangement of various components of the apparatus is shown in Fig. 3. Sodium vapour lamp is used as a source of monochromatic light, whose wavelength to be determined. Short focal length lens is used to render the incident rays parallel by placing the source of light at its focus. The light is reflected vertically downwards by a thin glass plate *P*, inclined at an angle of 45° to the horizontal.

The beam falls normally on the plane face of a plane-convex lens having large radius of curvature. The plano-convex lens is kept in contact with an optically flat glass plate. Few drops of liquid whose refractive index is to be determined are placed in between the lens and glass plate so that a uniform liquid film is formed. Newton's rings are thus formed due to interference between the rays reflected partly at the lower surface of the liquid film enclosed between the lens and the plate. These fringes can be seen through the travelling microscope adjusted vertically over the plate.

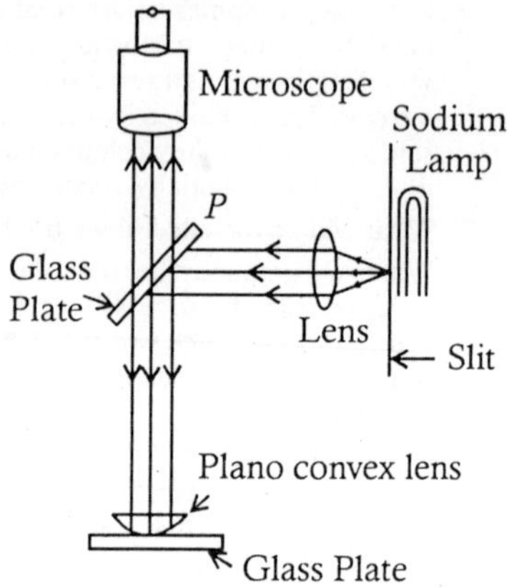

Fig. 3

The plane glass plate is placed on a blackned surface to avoid reflection from its lower surface. The whole system (inclined glass plate, plano-convex lens and plane glass plate) is placed in wooden structure blackened from inside.

Theory

The formation of Newton's rings is due to the interference of light waves reflected from the upper and lower surface of the liquid film enclosed between the lens and plane glass plate. The thickness of the air film at the point of contact is zero and increases as we move from the point of contact towards the periphery of the lens. The thickness of the air film will be uniform for all points laying on a circle with centre at the point of contact. Thus using reflected light,the central point will be dark surrounded by bright circles separated by dark rings.

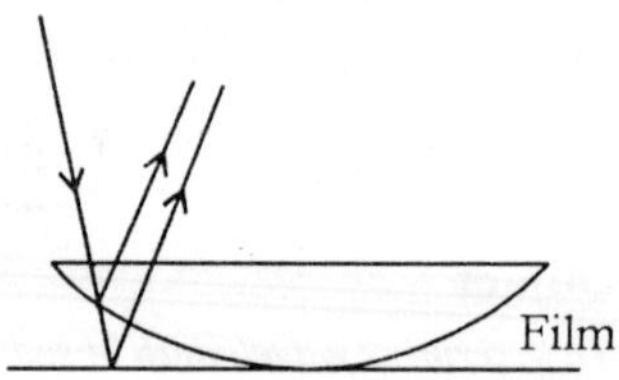

Fig. 4

If *R* is the radius of curvature of the lens and *t* the thickness of the air film at *A* or *B*. The necessary condition for bright fringes using reflected light is

$$2\mu t \cos\theta = (2n + 1)\lambda/2 \quad (1)$$

where $n = 0, 1, 2, \ldots$ Since θ is very small $\cos\theta = 1$. (θ is the angle of refraction)

$$\therefore \quad 2\mu t = (2n + 1)\lambda/2 \quad \text{(For air } \mu = 1) \quad (2)$$

and for dark fringes $2\mu t = (2n)\lambda/2 = n\lambda, \quad n = 0, 1, 2 \quad (3)$

From the geometry of Fig. 5

$$EB \times AE = EO \times ED$$

$$EB = AE = r \text{ (the radius of interference fringe).}$$

$$EO = BQ = t \text{ (thickness of the film)}$$

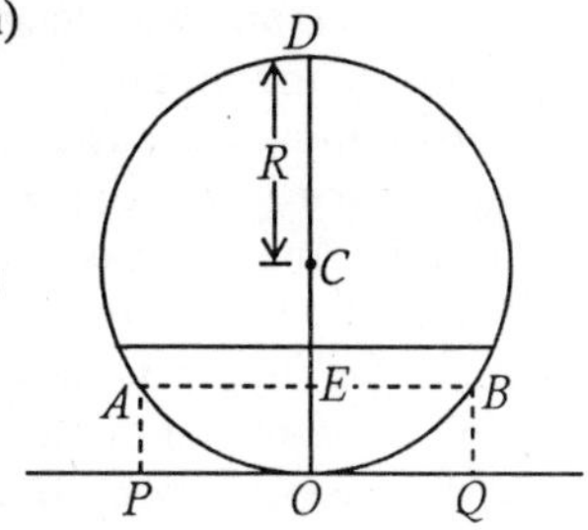

Fig. 5

$$\therefore \quad r_2 = t(2R - t) = 2Rt$$

Neglecting t^2 in comparision to $2Rt$

$$t = \frac{r^2}{2R} \tag{4}$$

$\therefore$ For bring fringe $2\mu \frac{r^2}{2R} = (2n+1)\lambda/2$

or

$$r = \sqrt{\frac{(2n+1)R\lambda}{2\mu}} \tag{5}$$

and for dark fringe

$$r = \sqrt{\frac{n\lambda R}{\mu}} \tag{6}$$

When $n = 0$, the radius of dark fringe is zero and bright fringe is $\sqrt{\lambda R}$. Therefore alternatively dark and bright fringes are produced with dark centre. Suppose the diameter of the nth dark fringe be D_n, then

$$D_n^2 = \frac{4n\lambda R}{\mu}$$

and the diameter of $(n + m)$th ring be D_{n+m}, then

$$\therefore \quad D_{n+m}^2 = \frac{4(n+m)\lambda R}{\mu}$$

$$\therefore \quad D_{n+m}^2 - D_n^2 = \frac{4m\lambda R}{\mu} \tag{7}$$

For bright fringes also

$$D_{n+m}^2 - D_n^2 = \frac{4m\lambda R}{\mu} \tag{8}$$

Procedure

Clean the lens surface as well as the glass plate thoroughly and arrange the apparatus as shown in Fig. 6. The focal length of the lens is pre-determined and is placed so that emergent beam is nearly parallel.

Adjust the source and inclined plate such that the central portion of the lens is illuminated. A spot with rings ground it can be seen by looking vertically above the lens. These are called Newton's rings which can easily be focussed by microscope.

For determining wavelength λ of light air film is taken and no liquid is used.

For determine the reflective index of liquid (μ) introduce some experimental liquid between the lens and glass plate.

Arrange the cross-wires of microscope with respect to the fringes as shown in Fig. 6 and measure the diameter of the fringes.

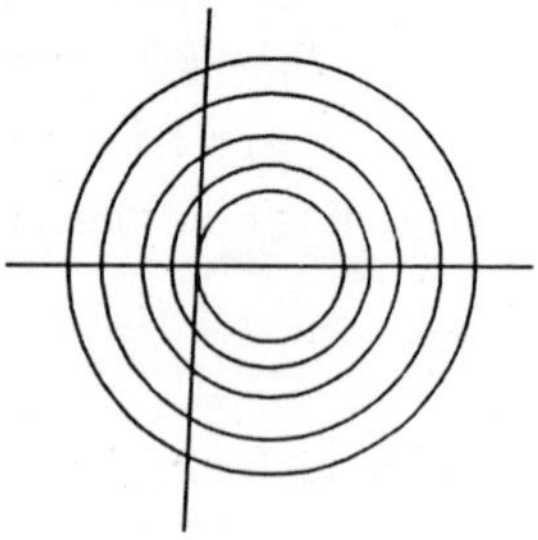

Fig. 6

Repeat the experiment with liquid film and find the $D'^2_{n+m} - D'^2_n$ for fringes in liquid film.

Determine the radius of the curved face of the lens in contact with the plate with the help of spherometer.

Observations

Determination of $D^2_{n+m} - D^2_n$ for air film

S. No.	*No. of ring*	*Microscope reading*		*Diameter D*	D^2	$D^2_{n+m} - D^2_n$ $(m = 7)$
		on L.H.S. of the ring	*on R.H.S of the ring*			
1 2 ⋮ 8						

Mean $D^2_{n+m} - D^2_n = \ldots \text{cm}^2$

Repeat the experiment with liquid film and find.

Mean $D'^2_{n+m} - D'^2_n = \ldots \text{cm}^2$

Determination of *R*

Least count of the spherometer = ... cm

Distance between the legs of spherometer *a* = ... cm

S. No.	*Readings of sphermeter when on*		*h (cm)*
	Plance surface (cm)	*Spherical (cm)*	

Mean *h*.

Calculations

(*i*) $R = \frac{a^2}{6h} + \frac{h}{2}$

(*ii*) $\lambda = \frac{D_{n+m}^2 - D_n^2}{4mR}$

(*iii*) $\mu = \frac{D_{n+m}^2 - D_n^2}{D'^2_{n+m} - D'^2_n}$

Substitue the values, and calculate R, λ and μ.

Result

The wavelength of the sodium light (λ) = ... Å.

Standard values of wavelength of the sodium light are 5890 Å and 5896 Å.

Refractive index of the liquid (μ) = ...

Precautions

1. The surface in contact must be thoroughly cleaned by xylene or rectified sprit.
2. The radius of the plano-convex lens should be fairly large in order to get large values of diameters. This will thus reduce the relative error in the measurement of D (diameter).
3. Short focal length lens should be used to get parallel rays.
4. The plate on which the plano-convex lens is placed with its curved surface downward should be optically plane, otherwise fringes will not be perfectly circular in shape.
5. The fringes should be focussed on the cross-wire without parallax.
6. The cross-wire should be fixed in the middle of the bright fringes. Use bright rings while measuring diameter as these provide a better contrast to the cross-wire.
7. Always move the micrometer screw in one direction so as to avoid back lash error.
8. Avoid measuring diameter of the first few rings, which are too thick and irregular in shape.
9. Since h is small (for determining R) as such it should be measured correctly.

Questions with Answers

1. What are Newton's rings and how they are formed?

Ans. When a plano-convex lens is placed on a glass plate, an air film of gradually increasing thickness is formed between them. When we allow to fall a monochromatic light normally on air film and viewed in reflected light, we observe alternate dark and bright Newton's rings. These rings are formed as a result of interference between the light waves reflected from the upper and lower surfaces of the air film enclosed between the convex surface of plano-convex lens and the plane glass plate.

2. Why the Newton's rings are circular?

Ans. In this experiment the air film is of varying thickness and a locus of constant illumination over the face of the film is a locus of path difference which is a circle with the point of contact of the lens and the glass plate as centre. This is why the fringes are circular.

3. What is the purpose of using plano-convex lens of large focal length?

Ans. Due to large radius of curvatures of plano-convex lens, the angle of wedge is very small and thus Newton's rings observed have large diameter. This helps to minimise the errors in measurments of diameter.

4. Why plane transparent glass plate is placed at 45° inclined plane?

Ans. This plate turns the light rays coming from an extended source to 90° and thus the light rays falls normally on plano-convex lens.

5. With the increase of the order of rings, rings get closer, why?

Ans. We have $D_n \propto \sqrt{n}$, i.e. the diameter of dark rings are proportional to square root of the natural number and $D'_n \propto \sqrt{2n-1}$, i.e. the diameter of bright rings are proportional to the square toot of odd natural number. Obviously, the diameter of the ring does not increase in the same proportion as the order of the ring increases. This causes the rings get closer and closer towards the ends.

6. Why an extended source of light is used in this experiment?

Ans. This helps to illuminate the entire surface of the lens. This is essential for the interference fringes to become circular rings.

7. What will happen if we use a point source or an illuminated slit instead of the extended source?

Ans. If we use a point source of light, only alternate dark and bright points will be seen. With an illuminated slit only a portion of the rings will be observed.

8. Why is the centre of the Newton's rings dark?

Ans. The actual path difference between the two interfering rays at the point of contact of plano-convex lens and plane glass plate is zero, but because one of the interfering rays is refleced from glass plate which is a denser medium, the effective path difference is $\lambda/2$ or a phase change of π. This means, the condition of minimum intensity at the centre is satisfied and obviously the central spot of the ring is dark.

10. When the Newton's rings are observed with transmitted light, will the central spot bright or dark?

Ans. Bright.

11. What will happen with the central spot when a liquid of refractive index μ greater than that of the lens and less then that of the glass plate is introduced between the lens and the glass plate?

Ans. In this case, the central spot becomes bright. Since the rays reflected from the lower surface of the lens (backed by denser medium) and that reflected from the upper surface of the glass plate will both suffer a phase change of π due to reflection.

12. Can you determine the refractive index of a liquid with this experiment?

Ans. Yes, we have $\mu = \dfrac{\left(D_{m+n}^2 - D_n^2\right)_{\text{air}}}{\left(D_{m+n}^2 - D_m^2\right)_{\text{liquid}}}$

13. What will happen to the Newton's rings, if the space between the lens and the plane glass plate is filled with a liquid of refractive index μ?

Ans. Newton's rings will contract in the ratio of $1/\sqrt{\mu}$.

14. What will happen if we replace the plano-convex lens by a transparent glass plate inclined at a small angle with the lower glass plate?

Ans. In this case one will observe the bright and dark straight line fringes in the field of view.

15. In this experiment, where the Newton's rings are formed?

Ans. Rings are formed on the air film between the lens and the plane glass plate.

16. It it desirable to measure the radius of curvature of the given lens by a spherometer in the usual manner?

Ans. No, the radius of curvature of the lens is very large and therefore the elevation of the central screw of the spherometer will be very small causing a huge error in the measurement.

17. Briefly explain the construction of microscope?

Ans. Travelling microscope used in this experiment consists of two convergent lenses. The lens towards an object is a convex lens of short focal length (objective lens) whereas the lens towards eye is called eye lens. The focal length of the objective lens should be smaller focal length than the eye lens. Both the objective lens and eye lens are mount at the two open ends of a cylindrical metal tube and the eye lens can be moved to and fro in the metal tube. Microscope helps to magnify the image of a very small object.

EXPERIMENT 4

Object

To determine the specific rotation of glucose / cane sugar solution using a polarimeter.

Theory

Let a plane polarized monochromatic light of wavelength λ pass through a column of solution of length l cm and containing m gms of an active substance per cm^3 at a given temperature. Then the plane of polarzation of the light is rotated through the angle θ given by

$$\theta = \frac{slm}{10}, \tag{1}$$

where s is the specific rotation of the solution. The quanity s is defined as the rotation produced by a column of solution of length 1 decimeter and containing 1`gm of the active substance per cm^3 of the solution under the experimental condition, i.e. at the temperature of the solution and for the wavelegth used.

If 100 cm^3 of the solution contains c gm of the active substance, the strength of the solution is c% and $m = c/100$. Hence,

$$\theta = \frac{slc}{1000}, \tag{2}$$

or

$$s = \frac{1000\theta}{lc}, \tag{3}$$

Equation (2) shows that for a given length l, the plot of θ versus c will be a straight line. This plot is known as the *calibration curve of the polarimeter* for the active solution.

This plot may be used to find the concentration c of an unknown solution of the same solute by measuring the rotation θ produced by it. From a set of

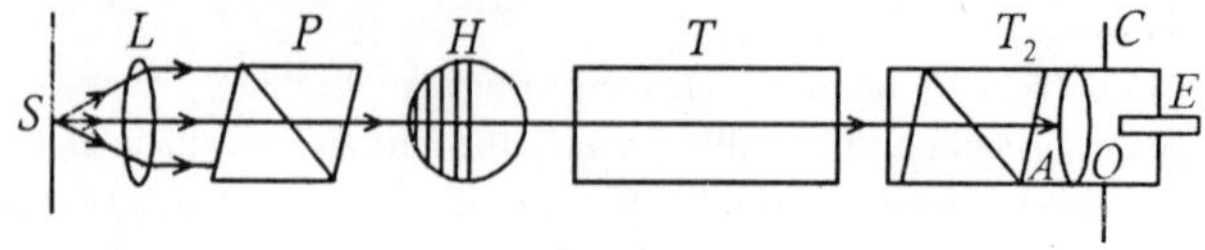

Fig. 7

S – Slit

L – Collimating lens

P – Polarizing nicol

H – Half-shade plate

T – Tube containing solution

A – Analysing nicol

O, E – Objective and eyepiece of the observing telescope.

T_2–Tube housing A and the observing telescope.

C – Cirular scale on which the vernier attached to T_2 rotates.

values of c and θ, and by measuring the length of the tube containing the solution, the specific rotation of the solution, s, can be determined from Eq. (3).

The amount of the solute, m', required to prepare the volume V cm^3 of the solution of concentration $c\%$ is given by

$$m' = \frac{cV}{100} \text{ gm}, \tag{4}$$

or

$$V' = \frac{100\,m'}{c} \text{ cm}^3, \tag{5}$$

From the solution of a known concentration $c\%$, a solution of a lower concentration $c'\%$ can be prepared by adding x'cm^3 of water to x cm^3 of the solution of concentration $c\%$, where

$$x' = \frac{c - c'}{c'} x \text{ cm}^3 \tag{6}$$

Equations (3), (4) and (6) are used as the working formulas for the experiment. If θ is measured in degrees, s will be obtained in degree cm^3 decimeter^{-1} gm^{-1}.

Apparatus

(*i*) Polarimeter, (*ii*) active subtstance (*iii*) balance and weights (*iv*) measuring cylinder, (*v*) flask (*vi*) distilled water, (*vii*) scale (*viii*) sodium light source, etc.

Procedure

Preparation of Solution of Known Concentrations

(*i*) First have a rough idea of the volume of the tube T (Fig. 7) of the polarimeter, which will contain the active solution. For it, fill the tube completely with water and measure the volume of that water by a measuring cylinder.

(*ii*) You are to prepare a solution of the highest concentration of a volume slightly (say 15 cm^3) greater than that of the tube T. Suppose you are to prepare a solution of concentration c% and volume V cm^3. Calculate the amount of solute, m' gm required to prepare this solution by Eq. (4). Now find the mass of a dry and empty flask by a balance. Indroduce certain amount of the solute into the flask and find the mass of the flask with the solute. From these two masses, find the mass m_1 of the solute introduced in the flask. Note that the amount of solute taken in the flask should be such that m_1 will be slightly greater than m'. Next introduce $V_1(= 100m_1/c)$ cm^3 of distilled water into the flask and shake the latter till the substance is dissolved completely in the water. Finally, filter the solution. This gives you a stock solution of the highest concentration c% and volume V_1 cm^3 which is slightly greater than V cm^3. Solutions of other known lower concentrations can be obtained from this stock solution of strength c%.

To Draw the Calibration Curve and Hence to Find the Concentration of Unknown Solution

(*i*) Find the vernier constant of the vernier attached to the analyser.

(*ii*) Focus the eye-piece of the telescope on the field of view of the polarizer.

(*iii*) Illuminate the slit S of the polarimeter by sodium light. Fill the tube T completely with distilled water taking care that no air bubbles exist inside the tube. Place the tube in its position. Rotate the tube T_2 of the polarimeter, which contains the analyzing nicol A and the vernier, till the two halves of the half-shade plate are equally bright. Take he readings of the two verniers for three independent settings.

(*iv*) Next remove the water from the tube T and wash it several times by taking a small amount of the stock solution of concentration c%. Then fill the tube completely with the solution. Take care that no air bubbles remain in the solution. On placing the tube in its position, you will see that the two halves of the half-shade plate to be unequally bright. Remote the tube T_2 in proper direction so that the two halves of the half-shade plate are again equally bright. Take the readings of the two verniers for three independent settings.

(v) Repeat operation (iv) for solutions of four other known concentrations and also for the solution of unknown concentration.

(vi) Plot the known concentrations along X-axis and the corresponding angles of rotation along Y-axis to get the calibration curve of the polarimeter for the solution of the active substance.

(vii) From the calibration curve determine the concentration coooresponding to the angle of rotation due to the unknown solution (of the same solute).

To Determine Specific Rotation of the Solution

(i) Measure the length between the two inner surfaces of the end plates of the tube T several times by a scale, and find the mean. This gives the value of l.

(ii) Take a point on the calibration curve sufficiently away from the origin so that the errors in reading the values of c and θ corresponding to this point may be neglibible. Note these values of c and θ and calculate the specific rotation of the active solution using Eq. (3).

Result

TABLE 1: PREPARATION OF SOLUTION OF HIGHEST CONCENTRATION c% BY VOLUME

Temperature of the solution = ...°C

Approx. inside vol. of the tube T (cm^3)	*Volume of the solution to be prepared,* V (cm^3)	*Mass of the solute required to prepare V* cm^3 *of soln. of concen. c%* $m'\left(=\frac{cV}{100}\right)$ (gm)	*Mass of the* *empty flask* (gm)	*Flask + solute* (gm)	*Mass of the solute actually taken,* m_1 (gm)	*Volume of water to be added to the flask to make the concentration c%* $v_1(=100\ m_1/c)$ (cm^3)
...	...	...	...+... +...+ ...+... =	...+... +...+ ...+... =	...	...

Precautions

1. In preparing the sugar solution, some sugar should be powdered with a pestel and mortar before taking the weight. The sugar should be dissolved completely in distilled water. The water should not be heated to assist dissolving. The solution should be filtered to make it dust free.

TABLE 2: CONVERSION OF SOLUTION OF CONCENTRATION c% INTO OTHER KNOWN LOWER CONCENTRATIONS

Concentrations required, c' (%)	*Concentration of the stock soln., c (%)*	*Volume of the stock soln. taken, x* (cm^3)	*Water to be added to the x* cm^3 *of the stock soln.,* $x'\left(=\frac{c-c'}{c'}x\right)$ (cm^3)	*Temperature of solution* (°C)
...	...	...	...	...
...	...	...	...	...

TABLE 3: VERNIER CONSTANT OF THE POLARIMETER

.... division of vernier = ... division of main scale

Value of smallest main scale division (l_1) (cm)	*Value of 1 vernier division* $\left(l_2 = \frac{n}{m}l_1\right)$(cm)	*Vernier constant* $V.C. = l_1 - l_2$ (cm)

TABLE 4: READINGS OF THE VERNIERS WHEN THE TUBE CONTAINS DISTILLED WATER

Wavelength of light used = ... Å

No. of observ-ations	*Readings of first vernier*				*Readings of second vernier*			
	Main scale (M)	*Vernier (V)*	*Total (T=M+V)*	*Mean*	*Main scale*	*Vernier (V)*	*Total (T=M+V)*	*Mean*
1	...	...	...	...	...	...	...	...
2	...	...	...	(To)	...	...	...	(To)
3	...	...	...	...	...	...	...	

TABLE 7: LENGTH OF THE TUBE *T* BETWEEN THE INNER SURFACES OF ITS END PLATES

No. of observations	*Length, l* (cm)	*Mean length, l* (cm)
1	...	...
2	...	...

TABLE 8: DETERMINATION OF THE SPECIFIC ROTATION OF THE SOLUTION

l (cm)	c (%)	θ (*degree*)	$s = \frac{1000\,\theta}{lc}$ (*degree*-cm^3 *decimeter*$^{-1}$ *gm*$^{-1}$)
...	...	...	...

TABLE 5: ANGLE OF ROTATION (θ) VERSUS CONCENTRATION (*c*) DATA

Wavelength of the light used = ... Å

Temp. of the solution (°C)	*Concentration of the solution, c (%)*	*Vernier No.*	*Main scale reading (M)*	*Vernier reding (V)*	*Total reading (T = M + V)*	*Mean reading*	*Angle of rotation from each vernier*	*Main angle of rotation, θ*
		1st	... ...	... ...	... ...	... (T_1)	... $(\theta_1 = T_1 \sim T_0)$	
		2nd	... ...	... ...	... ...	(T'_1)	$(\theta_1 = T'_1 \sim T_0)$	$\left(\theta = \frac{\theta_1 + \theta_2}{2}\right)$
		1st	... ...	... ...	... ...	...	...	
		2nd	... ...	... ...	... ...	... ...		

TABLE 6: DETERMINATION OF CONCENTRATION OF THE UNKNOWN SOLUTION FROM THE CALIBRATION CURVE

Temperature of the solution = ...°C; Wavelength of the light used = ... Å

Vernier No.	*Main scale reading (M) (%)*	*Vernier reading (V)*	*Total reading (T = M + V)*	*Mean reading*	*Angle of rotation form each vernier*	*Mean angle of rotation*	*% concentration from graph*
1st	... ...	... ...	... ...	... (T_1)	... $(\theta' = T_1 \sim T_o)$	$\left(\theta = \frac{\theta'_1 + \theta'_2}{2}\right)$	... (*c*)
2nd	... ...	... ...	... ...	(T')	$(\theta'_1 = T'_1 \sim T_o)$		

2. When the tube is filled with the solution care should be taken to remove air bubbles.
3. After placing the tube containing the solution in its position, it should be allowed to stand there for some time (at least five minutes) so that the temperature may become uniform, and then readings should be taken.
4. Since the specific rotation depends on temperature, the temperature of the solution of each concentration and also of the unknown solution should be noted while measuring the corresponding rotations. For this, a thermometer should be placed alongside the tube containing the solution and after suitable intervals of time, the temperature is to be noted.
5. While measuring the angle of rotations, the two halves of the half-shade plate should be equally bright. Failure to judge the equality of brightness will introduce errors. To minimize those errors, readings should be taken three times for each concentration for three independent settings of the analyser.

Questions with Answers

1. What is polarized light?

Ans. When all the rays of a beam of light wave have vibrations in one definite direction on a plane drawn perpendicular to the beam of light, the light is said to be polarized.

2. What are plane of polarisation and plane of vibration?

Ans. The plane containing the direction of vibration and the direction of propagation of light is called plane of vibration. The plane passing through the direction of propagation and containing no vibration is called plane of polarisation. We may note that plane of polarisation is always perpendicular to the plane of vibration.

3. What is an optically active substance?

Ans. A substance is said to be optically active, if the plane of polarisation of a polarized light is rotated in passing through it.

4. What are dextrorotatory and laevorotatory substances?

Ans. Dextrorotatory substances rotate the plane of polarisation of incident polarized light in clockwise direction whereas laevorotatory substances rotate in anticlockwise direction as seen by the observer facing the emergent light through it.

5. Mention few examples of dextrorotatory and laevorotatory substances.

Ans. Examples of dextrorotatory are: solution of comphor in alcohol, cane sugar, glucose, etc.

Example of Laevorotatory are: Fructose, Nicotine, Turpentine, etc.

6. On what factors the optical activity of a substance depend?

Ans. (*i*) Wavelength (*ii*) temperature.

7. How does the optical activity depend on wavelength?

Ans. The optical activity is inversely proportional to the square of the wavelength.

8. How does the optical activity depend on temperature?

Ans. The rotation for solids such as quartz, NaCl, etc. increases with the increase of temperature, while for liquids such as turpentine oil, essence of orange, etc. the rotation decreases with the increase in temperature. One can represent the rotation θ by $\theta = a + bt + ct^2$, where a, b and c are constants depend on the nature of the substance. We may note that constant b and c are negative for substances whose optical activity decreases with increase of temperature.

9. Mention practical application of the measurement of specific rotation.

Ans. One can use it for estimation of percentage of sugar in a given solution. This method is also use to determine the amount of sugar present in the urine of a diabetic patient.

10. What types of crystals do exihibit optical activity?

Ans. The crystals whose atoms of molecules are arranged on a spiral, exhibit the property of optical activity. When the spiral is left-handed, the rotation is left-handed whereas it is right-handed when the spiral is right handed.

11. What is molecular rotation?

Ans. Molecular rotation = specific rotation × molecular weight of the substance.

12. What is rotatory dispersion?

Ans. The phenomenon of the variation of rotation of the plane of polarization of the plane of polarized light with wavelength is called rotatory polarization.

13. What is the unit of specific rotation?

Ans. degree/decimeter/gm/cc.

14. What conclusion will you draw from the phenomenon of polarization of light?

Ans. Nature of light waves is transverse.

15. Will the plane of polarisation of a polarized light travelling through DNA rotate?

Ans. Since DNA solution is optically active and hence plane of polarization will rotate.

16. What is the difference between natural and magnetic rotations of plane of polarisation?

Ans. If the direction of light is reversed in natural rotation, the rotation is annulled, while in magnetic rotation, the effect is doubled. Moreover, the magnetic rotation is dependent on the orientation of the magnetic field, being a maximum when the light travels perpendicular to the magnetic field.

17. What is a polarimeter?

Ans. Polarimeter is used for the measurement of the angle of rotation of plane of polarisation rotated by an optically active substance.

The main parts of a polarimeter are polariser, analyser and polarimeter tube.

18. What is the function of polariser and analyser in this experiment?

Ans. Polariser changes an ordinary light into plane polarised light whereas analyser is used to analyse the light transmitted through the optically active substance and half shade on biquartz devices.

EXPERIMENT 5

Object

To determine the wavelength of prominent lines of mercury by plane diffraction grating.

Apparatus

A diffraction grating, spectrometer, mercury lamp, prism reading lens.

Formula

The wavelength λ of any spectral line can be calculated by the formula

$$(a + b)\sin\theta = n\lambda$$

or

$$\lambda = \frac{(a+b)\sin\theta}{n}$$

where $(a + b)$ = grating element, θ = angle of diffraction.

n = order of the spectrum.

Adjustment

(*a*) Before using the spectrometer, the following adjustments are made:

(*i*) The axis of the telescope and that of the collimator must intersect the principal vertical axis of rotation of the telescope.

(*ii*) Prism table should be levelled.

(*iii*) Telescope and collimator are adjusted for parallel light by Schuster's method.

(*b*) Grating should be normal to the axis of collimator. This adjustment is shown in Fig. 8

(*i*) Collimator and Telescope are arranged in a line and the image of the slit is focussed on the vertical cross wire. This reading is noted on both the verniers.

(*ii*) The telescope is shown rotated through 90°.

(*iii*) Mount the grating on the prism table and rotate the prism table so that the reflected image is seen on the vertical cross wire. Take the reading of the verniers.

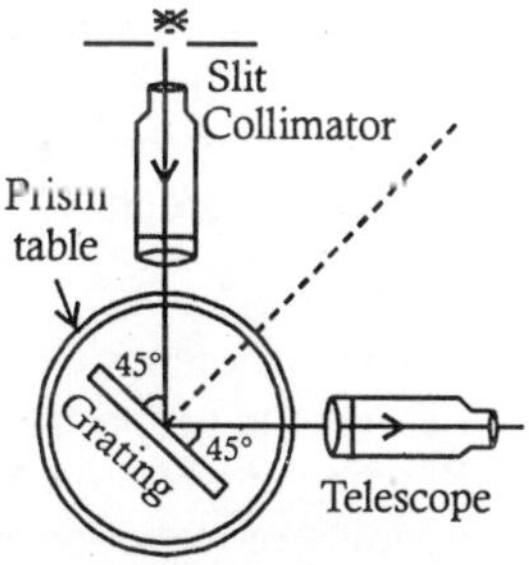

Fig. 8

(*iv*) Turn the prism table from this position through 45° or 135°. In this position the grating is normal to the incident beam.

(*c*) The slit should be adjusted parallel to the lines of the grating.

For this setting the slit is rotated in its own plane till the spectral lines become very sharp and bright.

Procedure for the Determination of Angles of Diffraction

The spectrum obtained in a grating is shown in Fig. 9

(*i*) Rotate the telescope to the left of the direct image and adjust the different spectral lines (violet, green and red) on the vertical cross wire for 1st order. Note dawn the readings of both the verniers.

(*ii*) Rotate the telescope further to obtain the second order spectrum and again adjust the spectral lines on the vertical cross wire and note the readings.

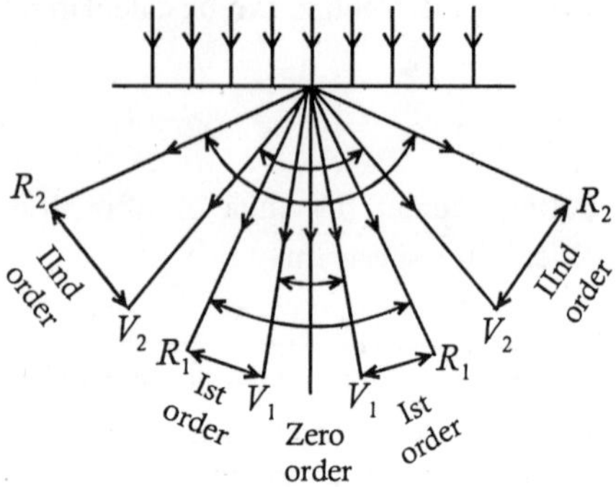

Fig. 9

(*iii*) Now rotate the telescope to the right of direct image and repeat the above procedure for first order as well as for second order.

(*iv*) Find out the difference of the same kind of verniers for the same spectral lines in the first order and in the second order. The angle is twice the angle of diffraction for that particular colour. Half of it will be the angle of diffraction.

(*v*) Find out the angle of diffractions for other colours in first and second orders.

Observations

Grating element $(a + b)$	= ... cm
Least count of spectrometer	= ... cm
Reading of telescope for direct image	= ...
Reading of telescope after rotating it through 90°	= ...
Reading of circular scale when reflected image is obtained on the cross wire	= ...
Reading after rotating the prism table through 45° or 135°	= ...

Calculations

Grating element $(a + b) = \dfrac{2.54}{N} = \ldots$ per cm,

where, N = No. of rulings per inch on the grating.

The wavelength of various spectral lines in the first order can be calculated by

$$\lambda = \frac{(a+b)\sin\theta}{n} = (a+b)\sin\theta$$

TABLE 1: DETERMINATION OF ANGLES OF DIFFRACTION

Order of spectrum	*Colour of light*	*Kind of vernier*	*Spectrum on left side*			*Spectrum on right side*			$2\theta = a + b$	*Mean* θ *Degrees*
			M.S. reading	*V.S. reading*	*Total* (*a*) *Degrees*	*M.S. reading*	*V.S. reading*	*Total* (*b*) *Degrees*		
First	Violet	V_1	...	...	...	...	...	...	...	...
		V_2	...	...	...	...	...	...	...	...
	Green	V_1	...	...	...	...	...	...	...	...
		V_2	...	...	...	...	...	...	...	...
	Red	V_1	...	...	...	...	...	...	...	...
		V_2	...	...	...	...	...	...	...	...
Second	Violet	V_1	...	...	...	...	...	...	...	...
		V_2	...	...	...	...	...	...	...	...
	Green	V_1	...	...	...	...	...	...	...	...
		V_2	...	...	...	...	...	....	...	...
	Red	V_1	...	...	...	...	...	...	...	...
		V_2	...	...	...	...	...	...	....	...

λ violet =...Å

Calculate λ for other spectral lines.

Wavelength in second order is given by

$$\lambda = \frac{(a+b)\sin\theta}{2}$$

λ violet =...Å

Calculate λ for other spectral lines.

Mean value of λ violet = ... Å.

Result

The wavelengths are given in the Table 2.

TABLE 2

Colour of spectral line	λ *Observed* (Å)	λ *Standard* (Å)
Violet I	...	...
Violet II	...	...
Blue	...	...
Green-Blue	...	...
Green	...	...
Yellow I	...	...
Yellow II	...	...
Red	...	...

Precautions

1. Before performing the experiment, the spectrometer should be adjusted.
2. Grating should not be touched by fingers.
3. Grating should be set normal to the incident light.
4. Both verniers should be read.
5. While taking observations, telescope and prism table should be kept fixed.

Questions with Answers

1. What is diffraction of light?

Ans. Light suffers some deviations from its rectilinear path while passing close to the edge of an opaque obstacle, or through a narrow slit; some light bends into the region of geometrical shadow. This phenomenon of bending of light around an obstacle is called diffraction.

2. When does diffraction of light become appreciable?

Ans. When the dimensions of the aperture or the obstacle in the way of the wave is comparable with the wavelength of the light.

3. What are the two classes of diffraction?

Ans. (*i*) Fresnel diffraction (*ii*) Fraunhofer diffraction.

4. Explain, how diffraction is a special kind of interference?

Ans. In interference, two primary waves from two coherent sources superpose and produce alternate bright and dark fringes. On the other hand, in diffraction phenomenon, secondary waves coming from the same wavefront superpose and produce alternate bright and dark fringes. Obviously diffraction is a special case of interference,

5. What is a plane transmission grating?

Ans. This is an optically plane parallel glass plate on which equidistant extremely close lines are made by ruling with diamond point. The space between the successive lines are transparent while the lines drawn are opaque to light.

6. What is reflection grating?

Ans. When the rulings are made on a polished metal surface, the grating becomes of reflection type.

7. What is a diffraction grating?

Ans. See *Q*. 5.

8. What is grating element?

Ans. It is the distance between the centres of any two successive ruled lines or transparent strips. If 'a' be the width of the slit and 'b' be the width of an opaque space, then the distance $(a + b)$ is the grating element.

9. What are corresponding points on the grating?

Ans. The points in the two slits separated by a distance equal to the grating element $(a + b)$ are said the corresponding points on the grating

10. What is dispersive power of a grating?

Ans. The rate of change of angle of diffraction with the change in wavelength of light used is called dispersive power of a grating. It gives an idea about the angular separation between the lines of spectrum produced by the grating.

11. What will be the effect of increasing the number of lines per cm on the grating?

Ans. When the number of lines per cm on the grating is increased, the angle of diffraction θ, for a particular order of grating spectrum also increases. This results in a reducing number of spectra separated by a large angle.

12. When even order spectra are absent in a grating spectra?

Ans. Even order spectra will be absent if the width of the clear space and ruled space is made equal in the grating.

13. What is the difference between a prism and grating spectrum?

Ans. In grating spectrum violet colour is least deviated while the deviation of red colour is maximum. The sequence of colours in prism spectrum is reverse than that of grating spectrum. The resolving power of grating spectrum is much greater than that of a prism. Moreover, the grating spectrum is purer than the prism spectrum.

14. Why we made light incident on the side of grating on which there are no rulings?

Ans. In order to avoid the refraction of diffracted light.

15. What will happen when the rulings of the gratings are not parallel and the distance between two successive rulings is not constant?

Ans. Near the actual spectral lines additional lines will appear which are termed as '*ghost lines*'.

16. What happens if ruled surface faces the collimator?

Ans. The rays will first diffract through the ruled surface and then the diffracted rays will undergo refraction passing through the glass plate. Obviously, the measured θ will not be due to diffraction alone.

17. Is there any secondary maxima formed?

Ans. Yes, however, they are not visible with a grating having large number of rulings.

18. How many orders of spectra are seen there?

Ans. First two orders of spectra are seen there.

19. Why we can not observe more than second order spectrum?

Ans. $(a + b)\sin\theta = n\lambda$

$$\sin\theta_{max} = 1 \qquad\qquad \therefore n = \frac{a+b}{\lambda}$$

Now, $a + b = \dfrac{2.54}{15000}$

$$n_{max} = \frac{a+b}{5893 \times 10^{-8}} = 2.87 < 3$$

Obviously, 3rd order spectrum is not possible.

20. How are commercial gratings prepared?

Ans. A commercial grating is prepared by pouring properly diluted cellulose acetate on the actual grating and drying it to a thin strong film. The film is detached from the original grating and then mounted between two glass plates. Thus a commercial grating is ready. This is also called a *replica grating*.

21. Define angular dispersive power of a grating.

Ans. Angular dispersive power of a grating is the rate of change of the angle of diffraction with the change in wavelength.

$$\frac{d\theta}{d\lambda} = \frac{nN}{\cos\theta} = \frac{n}{(b+c)\cos\theta}$$

where $(b + c)$ is grating element.

22. What is resolving power of a grating?

Ans. Resolving power is the capacity to separate the images of two closely spaced objects. R.P. = $\lambda/d\lambda$, i.e. two lines of wavelengths λ and $\lambda + d\lambda$ are just resolved or separated in the grating spectrum we have $\dfrac{\lambda}{d\lambda} = nN$, where N is the total number of rulings on the grating and n is the order number of the spectrum.

23. Mention uses of a diffraction grating.

Ans. (*i*) To study spectra and (*ii*) to measure wavelength.

EXPERIMENT 6

Object

To determine the resolving power of a telescope.

Apparatus

Telescope with a rectangular adjustable slit, a black cardboard with narrow white strips on it, travelling microscope and metre scale.

Formula

The theoretical and practical resolving powers are given by

$$\text{Theoretical resolving power} = \frac{\lambda}{a}$$

and $$\text{Practical resolving power} = \frac{d}{D}$$

where λ = mean wavelength of ligh employed,

a = width of the rectangular slit for just resolution of two objects,

d = separation of two objects,

D = distance of the objects from the objective of the telescope.

Hence $$\frac{\lambda}{a} = \frac{d}{D}$$

Theory

Rayleigh's Criterion of Resolution: According to Rayleigh's criterion, two equally bright sources can be just resolved by any optical system when their distance apart is such that in the diffraction pattern, the maximum due to one falls on the minimum due to the other.

Resolving Power of Telescope: The resolving power of a telescope may be defined as the inverse of the least angle subtended at the objective by two distance point objects which can be just distinguished as separate in its focal plane.

Let a beam of monochromatic light starting from a distant object O (not shown) be incident normally on a rectangular aperature AB fitted infront of the telescope objective. Let AQ represents the incident wavefront which is brought to a focus F and observed magnified by means of eyepiece. The intensity pattern at F is shown by thick curved line.

Consider again an object O' towards right of O whose pattern is formed towards left of the F. The pattern is formed at F' and shown by dotted curve. The wavefront due to the incident light is shown by AN. According to the Rayleigh criterion, the two objects can only be resolved when the maximum due to one falls on the minimum due to the other as shown in Fig. 10.

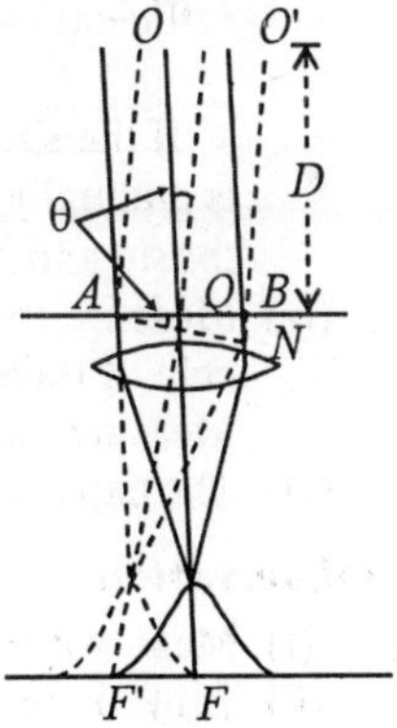

Fig. 10

As the aperture is rectangular the minimum due to one will fall on the maximum of the order when $QN = \lambda$. The angle between the two wavefronts is

$$\theta = \frac{AQ}{AN} = \frac{\lambda}{a}$$

where a is the aperture and θ is the angle subtended by the two objects OO' at the objective to telescope.

Again $$\theta = \frac{OO'}{D} = \frac{d}{D}$$

where d is the distance between two objects and D is the distance from the objective of the telescope.

Procedure

(*i*) Mount the telescope on a stand such that its axis lies horizontal and the rectangular lines marked on cardboard on the another stand such that they are vertical. Place the two stands at a suitable distance (say about 10 to 15 ft.) (Fig. 11).

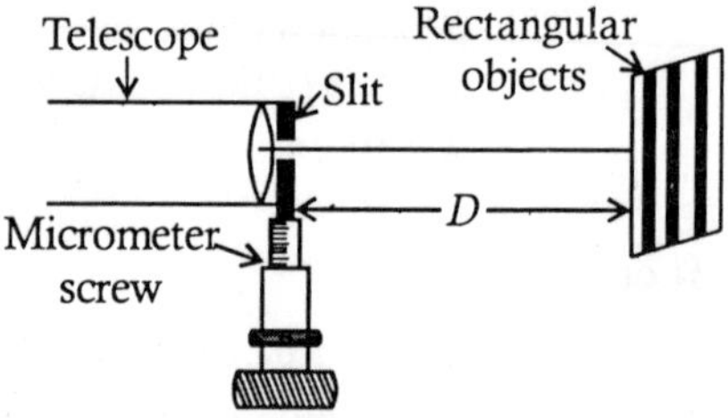

Fig. 11

(*ii*) Illuminate the objects with source of light. Now open the slit with the help of, micrometer screw and move the telescope in the horizontal direction such that the images of two vertical sources are seen in the field of view of the eyepiece.

(*iii*) Gradually reduce the width of the slit till the two images just cease to appear as two. Note down the reading of the micrometer. Again close the slit completely and note down the micrometer reading. The difference of the two readings gives the width of the slit (a) just sufficient to resolve the two images.

OR

If the slit is not provided with micrometer arrangement, the slit is gradually reduced till the two images cease to appear two. Take the slit and measure its width with the help of travelling microscope.

(*iv*) Measure the width of white or black rectangular object with the help of travelling microscope.

(*v*) Measure the distance between the object and the slit which gives D.

(*vi*) The experiment is repeated for different values of D.

Observation

(*i*) Mean value of $\lambda = 6000 \times 10^{-10}$ m

(*ii*) Table 1 for the width (a) of slit when micrometer arrangement is attached.

TABLE 1: L.C. OF SCREW = ...cm

S. No.	*Slit readings*						*Width of the slit* $a = (X-Y)$	*Theoretical resolving Power* (λ/a)	*Distance D*
	Solution ceases			*When slit is closed*					
	M.S. reading	*V.S. reading*	*Total X*	*M.S. reading*	*V.S. reading*	*Total Y*			
1	...	...	...	...	...	...	...	...	...
2	...	...	...	...	...	...	...	...	...

When micrometer arrangement is not used.

TABLE 2: TABLE FOR THE WIDTH OF SLIT (a)

L.C. of travelling microscope = ... cm

S.No.	*Distance D*	*Microscope reading*			λ/a
		M.S. reading	*V.S. reading*	*Total a*	
1	...	...	...	...	...
2	...	...	...	...	...
3	...	...	...	...	...

TABLE 3: TABLE FOR THE DISTANCE BETWEEN TWO OBJECTS (d)

L. C. of travelling microscope = ... cm

S.No.	*Distance D*	*Microscope reading (d)*			*Practical resolving power d/D*
		M.S. reading	*V.S. reading*	*Total d*	
1	...	...	...	...	...
2	...	...	...	...	...
3	...	...	...	...	...
4	...	...	...	...	...

Results

The theoretical and practical resolving powers of the telescope are as shown in the above tables.

Precautions

1. The axis of the telescope should be horizontal.
2. The rectangular objects drawn on the cardboard should be vertical.
3. Back lash error in the micrometer screw should be avoided.
4. The plane of the slit should be parallel to the objects.
5. The width a should be measured carefully.

6. The minimum width of slit for resolution should be adjusted very carefully.
7. The distance D should be measured from the slit of the telescope to the cardboard.

Questions with Answers

1. What is resolving power (R.P.) of a telescope?

Ans. R.P. is defined as the reciprocal of the angular separation that two distant object points must have, so that their images will appear just resolved. This is Rayleigh's criterion of R.P.

2. What do you understand by limit of resolution of a telescope?

Ans. Limit of resolution of a telescope is the minimum angle subtended at the objective of the telescope by two distant objects which can be separated by it. This angle is said to be the measure of the resolution by the telescope. We may note that smaller this value; higher will be R.P. of the telescope.

3. On what factors does R.P. of a telescope depend?

Ans. R.P. of a telescope is directly proportional to the wavelength and inversely proportional to the diameter of the aperature.

4. Why an objective with large diameter is fitted in the telescope?

Ans. To increase the resolving power of telescope.

5. How, one can increase the resolving power of a telescope?

Ans. By increasing the aperature of the lens used.

6. If the focal length of the objective is increased three times, how it will affect the R.P of the telescope?

Ans. There will be no effect on R.P. as R.P. is independent of focal length.

7. What way the objective of larger focal length of a telescope will affect?

Ans. Objective of larger focal length will only increase the magnifying power (M.P.) of telescope. This will not affect R.P.

8. Define M.P. of a telescope.

Ans. M.P. of a telescope is the ratio of the image subtended by the image at the eye to the angle subtended by the object at the eye when it is seen without the telescope.

$$\text{M.P.} = F/f$$

where F is the focal length of the objective and f is the focal length of the eye piece.

9. What is the limit of resolution of normal human eye?

Ans. About one minute.

10. What do you understand by 50" telescope?

Ans. The diameter of the telescope is 50".

11. What will be the limit of resolution of 40" telescope?

Ans.

$$d\theta = \frac{1.22\lambda}{a}$$

$$= \frac{1.22 \times 6000 \times 10^{-8} \times 180 \times 60 \times 60}{40 \times 2.54 \times 3.14}$$

$$= 0.15 \text{ s}$$

12. Why there is a difference between theoretical value and experimental value of R.P.'s.?

Ans. The difference between the two values is due to the different resolution limit for different eyes. We may note that there is enough arbitrariness about resolution limit.

13. Imagine that two close lines are just resolved by a telescope. Now the telescope is removed further from the lines. Will the two lines be still resolved?

Ans. No, with the increase of distance of the objective from the telescope, lines will not remain resolved. The angle subtended by the two lines at the objective in the first case was equal to the angle to minimum resolution. However, the angle subtended by the two lines becomes less than the angle for minimum deviation when the telescope is further taken away. This is why the telescope is not able to resolve the two lines in the second situation.

14. In Q. No. 13 what we can do to resolve the two lines again?

Ans. By increasing the width of the slit one can again get resolved these lines.

EXPERIMENT 7

Object

To determine the wavelength of sodium light with the help of Michelson interferometer.

Apparatus

Michelson's interferometer, sodium lamp, co-densing lens, a tin sheet having a fine pin-hole.

Description of the Instrument

It is built on an 'L' shaped cast-iron frame mounted on a heavy, rigid and firm stand on which two plane mirrors M_1 and M_2 silvered on their front surfaces (to avoid multiple internal reflections) are mounted vertically. The stand consists of a metal bed provided with a screw rod for movement of mirror M_1. It has a pitch of 1 mm and is fitted with a drum divided into 100 equal division. Rotation of the drum causes coarse motion of the screw while rotation of the precision head causes the slow motion. The precision head is also divided in 100 equal parts and one complete rotation of the head causes the rotation of the drum by one division. Thus one division of the slow motion head corresponds to $\frac{1}{100 \times 100} = 10^{-4}$ mm = 10^{-7} m which is least count of the screw rod.

The second mirror M_2 is fixed at the end of a metal arm mounted at the end of the bed at right angles to it. The position of the mirror M_2 along the bed can be altered by rotating other screw at its back. The arm carrying M_2 also carries two glass plates G_1 and G_2 which are cut from some optically worked plane parallel plate of glass. The rear (back) side of plate G is semi-silvered so that one-half of the incident light is reflected and other half is transmitted through it, G_2 carries a graduated screw for changing its inclination. Both plates are mounted in two separate frames and fixed in vertical planes parallel to each other and at an angle 45° to the arm. Each of the mirrors M_1 and M_2 is provided with a set of three fine screws with special screws with speceial arrangement of springs for adjusting the mirrors exactly perpendicular to each other.

Working of the Apparatus

Light from an extended source (monochromatic) S, is rendered parallel by condensing lens L and incident on semi-silvered plate G_1. Thus SO is one of the rays of light beam (Fig. 12). It is refracted in the plate G_1 to the point A, where it is partially reflected and partially transmitted. Thus the ray SO is amplitude divided into two rays of equal intensity 1 and 2. These rays proceed towards M_1 and M_2 respectively; are reflected normally from their front silvered surfaces, and reunite at the point A of the plate G_1. Now at A, a part of the amplitudes of the ray from M_2 is reflected while a part of that from M_1 is transmitted along A_1 and enter a short-focus telescope T. Since the rays enterring the telescope are derived from the same incident ray, they are coherent and hence in a position to produce interference. The intereference fringes can be seen in the telescope.

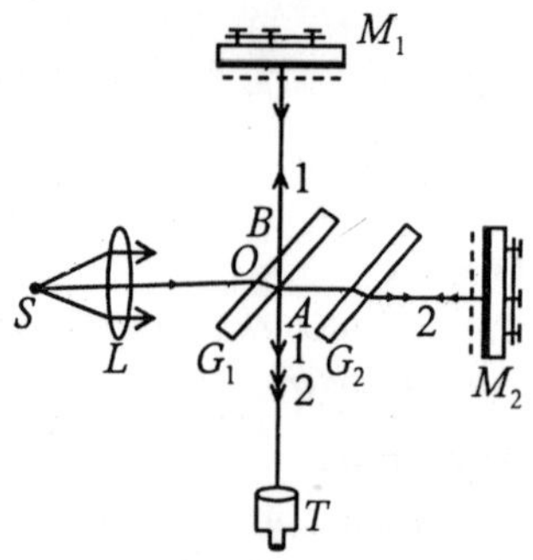

Fig. 12

Function of the Plate G_2

After the partial reflection and transmission at A; the ray 1 travels through the glass plate G_1 twice while ray 2 does not do so even once. Thus in the absence of G_2, the paths of rays 1 and 2 in glass are not equal. To equalise these paths the plate G_2 which has the same thickness as G_1 is placed parallel to G_1, G_2 is called the "*Compensating Plate*".

Form of Fringes

The form of fringes depends on the inclination of M_1 and M_2. Let M_2' be the image of M_2 formed by reflection at the semi-silvered surface of G_1 so that $AM_2' - OM_2$. The intereference fringes may be regarded to be formed by light reflected from the surface M_1 and M_2 respectively. Thus arrangment is equivalent to an air-film enclosed between the reflecting surface M_1 and M'_2.

When M_1 is exactly perpendicular to M_1, the film M, M_2' is of uniform thickness and we obtain circular fringes localised at infinity.

Theory

The interferometer is adjusted for circular fringes and a bright spot is obtained at the centre of the field of view. Now if d is the thickness of the air-film enclosed between M_1 and M'_2 and n the order of the fringe at centre, then

$$2d\cos\theta = n\lambda \tag{1}$$

But at the centre $\theta = 0$, $\cos\theta = +1$

$$2d = n\lambda \tag{2}$$

Now if M_1 is moved away from M'_2 by $\lambda/2$ then $2d$ increases by λ. Therefore $(n + 1)$th order fringe replaces nth other fringe. Thus each time M_1 moves through a distance $d/2$, next bright fringe appears at the centre. If on moving M_1 through a distance x, N mew fringes appear at the centre of the field, then

$$x = N\lambda / 2 \text{ or } \lambda = \frac{2x}{N} \tag{3}$$

Determination of $\Delta\lambda$

When the source of light has two wavelengths λ_1 and λ_2 very close together, then each wavelength produces its own system of rings. Let $\lambda_1 > \lambda_2$, when thickness of the air-film between M_1 and M'_2 is small, the rings due to λ_1 and λ_2 almost coincide since λ_1 and λ_2 are nearly equal. If now the mirror M_1 be moved away, then due to different spacing between the rings of λ_1 and λ_2, the rings of λ_1 will gradually separate from those λ_2. When the thickness of the air-film becomes such that a dark ring of λ_1 concides with a bright ring of λ_2 the field of view will have uniform intensity and no fringes will be observed. This is called *dissonance.*

The mirror M_1 is moved further away through a distance x (say) until the rings, after becoming most distinct, once again become most indistinct. Clearly, during this movement n fringes of λ_1 and $(n + 1)$ fringes of λ_2 have appeared at the centre (because then the dark fringe of λ_1 will again coincide with the bright rings of λ_2). Now since the movement of the mirror M_2 by $\lambda/2$ results in the appearance of new fringe at the centre, we have

$$x = n\frac{\lambda_1}{2} = (n+1)\frac{\lambda_2}{2}$$

or
$$\frac{2x}{\lambda_1} = n \text{ and } \frac{2x}{\lambda_2} = n+1$$

$\therefore$
$$\frac{2x}{\lambda_2} - \frac{2x}{\lambda_1} = 1$$

or
$$\Delta\lambda = \lambda_1 - \lambda_2 = \frac{\lambda_1\lambda_2}{2x} = \frac{\lambda^2}{2x} \tag{4}$$

where $\lambda^2 = \lambda_1\lambda_2$. Here λ is taken as mean of λ_1 and λ_2.

Adjustment of the Instrument

In order to obtain the circular fringes the image of M_2 in G_1, i.e. M'_2 must be exactly parallel to real mirror M_1. For this following adjustments are made.

The distances of mirror M_1 and M_2 from G_1 are kept nearly equal within a millimeter. Light from source is rendered parallel by condensing lens L. A plane paper with pin hole is placed between lens and G_1 such that centres of mirror M_2, lens L and the pin hole lie in the same horizontal line. On looking into instrument through G_1, four images R_1, R_2, R_3, R_4 are seen in the field of view corresponding to four emergent rays. Two images R_1 and R_2 are brighter and are formed by light undergoing partial reflections at the semi-silvered surface of G_1 and reflections at M_1 and M_2. The other two images R_3 and R_4 are fainter and are formed by partial reflections at the unsilvered surface of G_1 and reflctions at M_1, and M_2. The brighter images are then made to coincide with fainter images by adjusting the screws behind the mirror M_2.

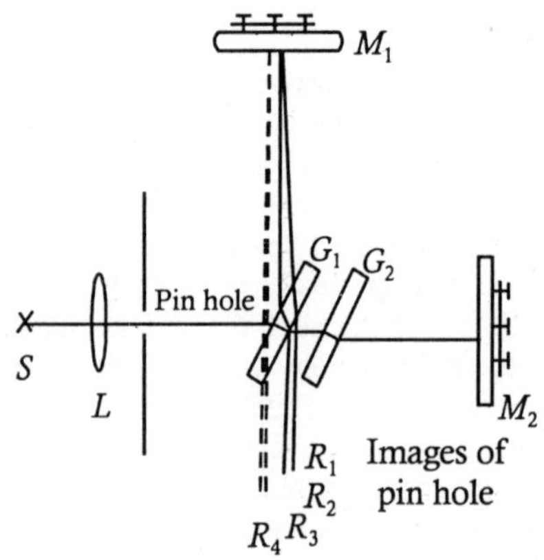

Fig. 13

The pin hole is removed and localised fringes appear. The screws behind mirror M_2 are adjusted till the spacing between these fringes becomes maximum and the centre of the fringe system comes in field of view. If linear or lateral movement of eye do not cause the fringe system to diverge or converge means the adjustment is perfect and the fringe system is situated at infinity.

Procedure: Deteremination of Wavelength

The interferometer is adjusted to obtain circular fringes which are observed through the telescope and a pinhole. The position of M_1 is then so adjusted that a bright spot appears at the centre of the field of view. The micrometer screw reading is then noted.

The mirror M_1 is moved away so that a good number of fringes (say 50) appear at the centre of the field. The micrometer screw reading is again noted.

This procedure is repeated a number of times.

Determination of $\Delta\lambda$: The interferometer is adjusted to produce circular fringes. The sodium yellow light consists of two wavelengths very close together and each wavelength gives its own set of fringes which operlap.

The position of mirror M is adjusted so that the combined fringe pattern shows maximum indistinctness. The micrometer screw reading is noted.

The mirror M_1 is further moved away until we get the next position of maximum indistinctness. The position of mirror M_1 is again noted.

The procedure is repeated for a number of consecutive position of maximum indistinctness.

Observation

Least count of the instrument = 10^{-7} m.

(*i*) For determination of wavelength λ.

S.No.	*No. of fringes*	*Position of the mirror M_1*	*Diff. of 250 fringes*
1	0		
2	50		
3	100		6-1
4	150		7-2
5	200		8-3
6	250		9-4
7	300		10-5

Mean difference x = ...

Similar readings are taken for other sets also. At least 5 sets are taken.

(*ii*) For determination of $\Delta\lambda$.

S.No.	*Position of mirror M_1 for dissonance*	*Difference*
1		(2 - 1)
2		(3 - 2)
3		(4 - 3)
4		(5 - 4)
5		(6 - 5)

Mean difference x = ...

Calculations

Calculate $\lambda = \dfrac{2x}{N}$

and $\Delta\lambda = \dfrac{\lambda^2}{2x}$

Here use λ as calculated above,

Result

The average wavelength of sodium yellow light is = ...
Standard value is 5893 Å.
The difference in wavelengths of the two *D*-lines $\Delta\lambda$ = ...
Standard value 6Å. (5896 Å – 5890 Å = 6 Å)

Precautions

1. The glass plates G_1, G_2 and the mirrors M_1 and M_2 should not be touched.
2. During the experiment mechanical disturbance should not be there.
3. The micrometer screw should be handled carefully.
4. The screw behind the mirrors should be moved slowly and slightly.

The total rotation should not exceed 1/8 of a rotation.
The centre of the fringe system must be at the centre of the cross wire or pin hole.

At the position of maximum indistinctness the fringes should almost disappear.

This method is the most accurate one for the determination of the wavelength of a monochromatic source but however is not suitable for determination of $\Delta\lambda$ (the difference between wavelength) because the position of maximum indistinctness can not be located correctly. For this purpose Fabery Perot interferometer is recommended.

We must note that if, there is even a slight mishandling of the screws behind the mirrors, it become difficult to adjust the apparatus to a good degree of accuracy.

Questions with Answers

1. What is an interferometer?

Ans. It is a device used to determine the wavelength of light. This utilizes the phenomenon of interference.

2. Why is a semi-silvered glass plate used?

Ans. This plate divides the amplitude of incident wave in two parts by partial reflection and partial refraction.

3. Explain the function of glass plate G_2 (compensating plate).

Ans. In this experiment the light wave is reflected thrice at G_1 through its thickness prior to superposition, while the refracted wave gets reflected only once. Obviously, an additional optical path difference is produced in reflected wave. In order to compensate this additional path difference an unsilvered glass plate G_2, identical to the plate G, is introduced parallel to G_1 in the path of the reflected wave. This is why G_2 is also called the compensating plate.

4. How interference fringes are produced by Michelson's interferometer?

Ans. By using the phenomena of partial reflection and partial refraction and by division of amplitude of incident wave, interference fringes are produced.

5. What types of two mirrors are used in this interferometer?

Ans. Purely optically plane and highly silver polished plane mirrors.

6. Where is the thin air film formed?

Ans. It is formed between the mirror M_1 and the image of mirror M_2.

7. How the glass plates G_1 and G_2 are mounted? What types of these plates are?

Ans. G_1 and G_2 plates are optically plane and of same thickness and same material. These plates are mounted parallel to each other at an angle of 45° with the two mirrors. Plate G_1 is made semisilvered at the back surface.

8. What types of fringes are obtained in this experiment?

Ans. Depending upon the path difference between the two waves and angle between the two mirrors, the fringes may be straight, circular or of any other shape.

9. How the mirrors M_1 and M_2 are mounted?

Ans. The mirror M_1 is mounted on a carriage which can be moved backward or forward and the mirror M_2 is fixed.

10. Why the mirrors M_1 and M_2 are silvered on the front surfaces?

Ans. To avoid multiple internal reflections.

11. How can one get circular fringes?

Ans. When two mirrors are exactly perpendicular to each other, the fringes obtained are circular. For this purpose, the screws provided at the back of the mirror M_2 are adjusted. We may note that the circular fringes are formed at infinity and that is why a telescope is used.

12. What are localized fringes?

Ans. Localized fringes are produced when the two mirrors M_1 and M_2 are not exactly perpendicular to each other. In such a situation either straight or parabolic fringes are observed.

13. How many fringes will appear or disappear, when the mirror is moved through a distance $\lambda/2$?

Ans. One fringe.

14. What will happen if we use white light source?

Ans. With white light source, we observe a central white (achromatic) fringe and some coloured fringes symmetrically situated on both sides of the central fringe.

15. What will happen if there is no compensating plate and we get localized fringes with white light?

Ans. In the absence of the compensating plate due to additional path difference (large) the whole fringe system may get displaced out of the field of view.

16. How is the above difficulty overcomed with the compensating plate?

Ans. The compansating plate nullifies the additional path difference and fringe pattern appears in the field of view.

17. What is the use of an extended source of light in this experiment?

Ans. The rays from each point of the extended source of light are divided and recombination takes place between rays which originally belonged to the same ray. Then the rays fall on the plate G_1 in the same direction. All rays suffer the same phase change and come out parallel so that the telescope able to focus them together in its focal plane.

18. What is the shape of air film?

Ans. The air film is of uniform thickness, if the mirrors M_1 and M_2 are exactly perpendicular to each other. The air film will be of wedge shaped if M_1 and M_2 are not perpendicular.

19. What is the difference between Michelson's circular fringes and Newton's rings?

Ans. The difference between the two are: (*i*) With the increase of radii of circular fringes, the order of interference decreases, while in Newton's rings, which are superficially similar, the order of interference increases with the increase in radii. (*ii*) The Newton's rings are loci of constant inclination whereas Michelson's circular fringes are loci of equal optical thickness. (*iii*) Michelson's circular fringes are situated at infinity whereas the Newton's rings are localized in the film.

However, when we increase their radii, rings in both the systems get closer.

20. What measurments can be made with Michelson's interferometer?

Ans. (*i*) Measurement of wavelength of light, (*ii*) determination of refractive index, (*iii*) thickness of thin transparent sheet and (*iv*) determination of difference between two near wavelengths e.g. Sodium, D_1 and D_2 lines.

EXPERIMENT 8

Object

To study elliptically-polarized light by means of a photoelectric cell.

Apparatus

Photoelectric cell, light source, doubly-refracting plate, a pair of polaroids (or nicols), sensitive dead-beat galvanometer, lamp and scale arrangements, slit, optical bench with suitable uprights, etc.

Photoelectric Cell

A photoelectric cell is a device which converts light energy into electrical energy. They are of different types; photo-emissive, photo-voltaic and photo-conductive, the first two being more common.

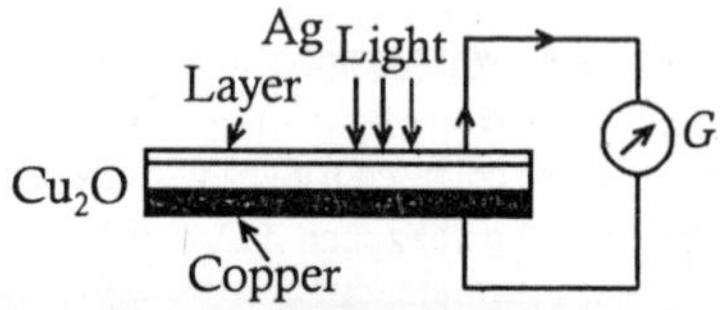

Fig. 14

A photo-voltaic cell (Fig. 14) consists of a copper plate on which a semi-conducting layer of cuprous oxide (Cu_2O) is formed which, in turn, is coated with a thin semi-transparent layer of silver (Ag). When light falls on the cell, it penetrates the Ag layer and photo-electrons are ejected from the Cu_2O layer. These electrons move towards the silver layer which then becomes negatively charged with respect to the copper plate. Hence a potential difference is developed between the silver layer and the copper-plate and photo-electrons flow in the external circuit. The strength of photo current is directly proportional to the intensity of light incident on the cell. This cell requires no external source of e.m.f. (battery).

Theory

When a beam of plane-polarised light is sent through a thin doubly-refracting plate, it breaks up into two coherent plane-polarised components vibrating at right angles to each other with a certain phase difference. The emergent light is the resultant of these two components and is in general, elliptically polarised. The constants of the elliptical polarisation, i.e., the phase difference between the two components, and the position and ratio of the axes of the ellipse, depend upon the thickness and optical properties of the doubly refracting plate used, and upon the angle which the plane of vibration of the incident light makes with the principal plane of the plate. This angle should be incident light makes with the principal plane of the plate. This angle should be different from 0, $\pi/4$ and $\pi/2$, because angles 0, and $\pi/2$ causes it circularly polarised. When the given elliptically-polarised light is passed through an analyser (polaroid or nicol) rotating about the direction of light as axis, a variation in intensity is observed but the minimum intensity is not zero. The intensity is a maximum when the principal section of the analyser is parallel to the major axis. If the light transmitted through the rotating analyser be allowed to fall on a Photo-electric cell, it will be converted into corresponding fluctuating current which will be a measure of the light intensity for various orientations of the analyser. From a suitable plot of the results, the ratio of the axes of the elliptical vibrations may be determined.

Let the major and minor axes of the given elliptical vibration be along the coordinate axes OX and OY (Fig. 15). Let a and b be the semi-major and semi-minor axes of the ellipse. Then the equation of the ellipse is

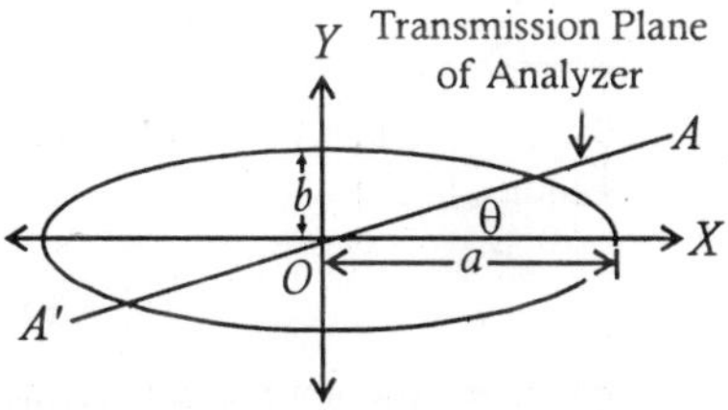

Fig. 15

$$\frac{x^2}{a^2} + \frac{y^2}{b^2} = 1$$

Let θ the angle between the major axis and the plane of transmission (AA') for the analyser. The amplitudes of the disturbances along OA due to the components along OX and OY will be $a\cos\theta$ and $b\sin\theta$ respectively, with a phase difference of $\pi/2$ between them. The intensity of the light transmitted by the analyser is, therefore, given by

$$I = a^2\cos^2\theta + b^2\sin^2\theta + 2(a\cos\theta)(b\sin\theta)\cos\pi/2$$

$$= a^2\cos^2\theta + b^2\sin^2\theta$$

$$=(a^2 - b^2)\cos^2\theta + b^2 \quad (1)$$

$$\left.\begin{array}{l}\text{Hence, for } \theta = 0° \text{ and } 180°;\ I = I_{max} = a^2 \\ \text{and for } \theta = 90° \text{ and } 270°;\ I = I_{max} = b^2\end{array}\right\} \quad (2)$$

Thus during a complete rotation of the analyser the intensity of the transmitted light passes through two maxima and two minima.

Equation (1) shows a graph of I against $\cos^2\theta$ will be a straight line, the slope of which with the $\cos^2\theta$–axis will give $(a^2 - b^2)$. Also, a^2 can be found by observing the maximum intensity (Eq. 2). Hence b^2 and the ratio a/b can be obtained (The intensity I is measured in terms of the galvanometer deflection).

Procedure

The experiment is arranged as shown in Fig. 16. S is an electric bulb carrying an ammeter and a rheostat in the circuit of its filament which is fed either by a battery of cells or by mains through a suitable transformer.

S_1 and slit and L is a converging lens, P_1 and P_2 are the polarising and analysing polaroids (or nicols) and Q is a doubly-refracting plate $\left(\text{a mica sheet or a } \frac{\lambda}{4} \text{ plate}\right)$. Behind P_2 is the photo-cell connected to a sensitive galvanometer which carries a damping key across it. All these pieces of apparatus are mounted in suitable uprights placed on an optical bench. P_1, P_2 and Q can be roated in their own planes about the axis of the light beam and their positions can be read on circular scales.

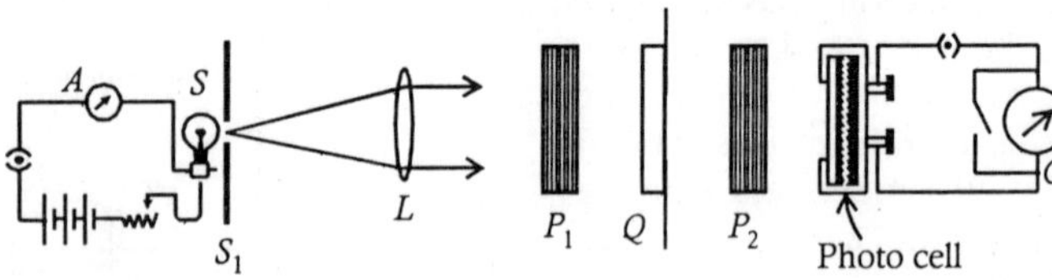

Fig. 16

Light from the source S is rendered parallel by the lens L and made plane-polarised by P_1. The doubly-refracting plate Q makes it elliptically-polarised. It is then transmitted through the analyser P_2, the intensity of transmission being different for different orientations of P_2. The transmitted light is received by the photo-cell which converts it into a current giving a deflection in the galvanometer G. The deflection is obviously different for different orientation of P_2. The procedure of the experiment is as follows:

1. First of all, the photo-cell is removed from the optical bench and the source S is switched on. The polariser P_1 and the doubly-refracting plate Q are adjusted at a relative orientation such that on looking through the analyser P_2, which is being rotated about the axis of light, the intensity varies between a maximum and minimum but is never zero. This sensures that the light emerging from Q is

elliptically-polarised. *(This adjustment of P_1 and Q is not disturbed for the entire set of observations.) The source S is switched off.

2. The galvanometer is levelled and its coil is released free. The lamp and scale arrangement is adjusted so that a sharp spot of light is obtained on the scale. The initial position of the spot, when at rest, is noted on the scale. The position is adjusted to be 0.0 cm.
3. The photo-cell is placed on the optical bench. The analyser P_2 is oriented such that its pointer reads 0° on the circular scale. The source S is switched on, and the deflection of the spot of light obtained on the scale is noted. The source is then switched off and the spot of light is brought to rest by pressing the damping key.
4. The analyser P_2 is rotated in steps of 10° and the corresponding deflections are noted in the same way as above. This is done until P_2 is given a full rotation of 360°.
5. Different sets of observations are taken by altering the relative orientation of the polariser P_1 and the doubly-refracting plate Q; but in each case it is ensured that the light emerging form Q is elliptically-polarised.

Observations and Calculations

I Set : Reading of the ammeter in series with the source =...
Relative orientation of polariser and doubly-refracting plate =... °
Initial reading of the spot of light on the scale = 0.0 cm.

TABLE 1

S. No.	*Orientation of the analyser* α	*Intensity I in terms of gal. deflection* (cm)	*S. No.*	*Orientation of the analyeser* α	*Intensity I in terms of gal. deflection* (cm)
1	0°	...	20	190°	...
2	10°	...	21	200°	...
3	20°	...	22	210°	...
...	...	...	...	...	...
...	...	...	...	...	...
19	180°	...	37	360°	...

Similar tables are made to the other sets of observations.

We plot the measured intensity of light I (in terms of galvanometer deflection for various orientations (angle α) of the analyser throughout a complete rotation with respect to an arbitrarily chosen reference line. The

* If the light emerging from Q were circularly-polarised, then on looking through the rotating analyser P_2 there would have been no variation in intensity. If it were plane-polarised, then there would have been an intensity variation with zero minimum.

shape of the curve so obtained resembles the figure of eight (Fig. 17) The intensities (deflections in cm) are represented by the radius vectors like OA and the orientations by the angle α with respect to the reference line. The maximum value of the radius vector defines $\theta = 0°$ (or 180°) and the minimum value defines $\theta = 90°$ (or 270°) in accordance to Eq. (2).

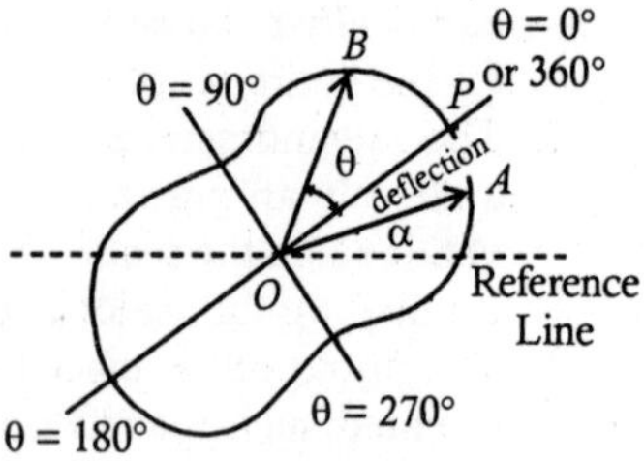

Fig. 17

Now we draw radius vectors (like OB) at angle $\theta = 0°, 10°, 20°, \ldots, 90°$ with respect to the longest vector (OP) and measure their corresponding to the orientations θ of the analyser with respect to the major axis of the elliptical vibration. These values are tabulated as below:

TABLE 2

S.No.	θ (degrees)	cos θ	$\cos^2\theta$	I (cm)
1	0	1-0000	1-0000	...
2	10	0-9848	0-9701	...
3	20	0-9397	0-8831	...
4	30	0-8660	0-7499	...
5	40	0-7660	0-5866	...
6	50	0-6428	0-4130	...
7	60	0-5000	0-2560	...
8	70	0-3420	0-1169	...
9	80	0-1736	0-0301	...
10	90	0-0000	0-0000	...

Now draw a graph between I and $\cos^2\theta$, which is a straight line (Fig. 18).

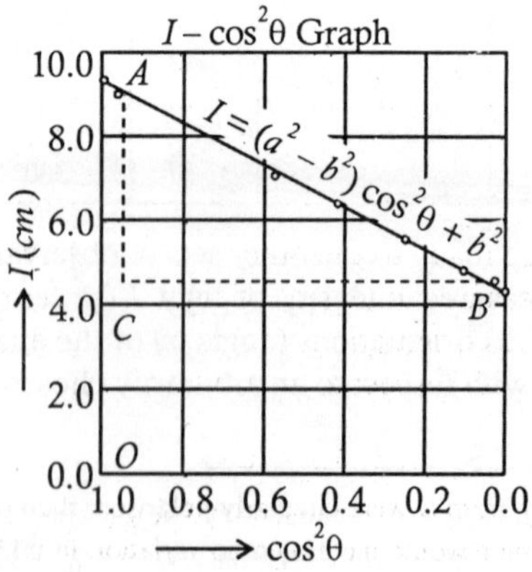

Fig. 18

From this graph, one obtains

$$a^2 = I_{max} \text{ (for } \theta = 0, \cos^2\theta = 1.0000)$$
$$= OP = \ldots$$
$$(a^2 - b^2) = \text{slope of the line} = \frac{AC}{CB} = \ldots$$
$$b^2 = \ldots$$

Then $$\frac{a^2}{b^2} = \ldots$$

and $$\frac{a}{b} = \ldots$$

Similarly the ratios $\frac{a}{b}$ are obtained from the other sets of observations.

Results

TABLE 3

S.No.	*Relative orientation of polariser and doubly-refracting plate (in degrees)*	*a/b*
1	...	...
2	...	...
3	...	...

The orientation of *I-x* plot (figure of eight) and the ratio of the axes of ellipse *a/b* are different for different sets of observations. This shows that the position and ratio of the axes of the elliptical vibration depend upon the angle which the plane of vibration of the incident plane-polarised light makes with the principal plane of the doubly-refracting plate.

Precautions

1. The optical-bench carrying the various pieces of apparatus should be kept covered by a black cloth so that no external light reaches the photo-cell.
2. The photo-cell should not be exposed to the light under investigation unnecessarily. The source should be switched on (while the photo-cell is on the optical bench) only when the reading is to be taken.
3. The current in the filament of the bulb (source) should be maintained constant by adjusting the rheostat in its circuit, otherwise the intensity of the incident light will vary, and this variation will be super-imposed on the variation in the intensity of light transmitted through the rotating analyser.
4. The galvanometer used should have sensitivity of the order of 10^{-8} amp. per mm at a scale distance of a metre, so that deflections in

the range 4 -12 cm are obtained. It should be levelled so that the coil has a free movement between the pole-pieces.

5. The galvanometer coil should be brought to rest by pressing the damping key after each observation. It should be checked that the initial position of the spot of light on the scale is always 0.0 cm.

Questions with Answers

1. Explain the functions of two polaroids in this experiment.

Ans. Polaroid P_1 acts as a *polarizer.* When light from the source, after passing through the slit and the lens, is allowed to fall on polaroid P_1, it makes its plane polarized by transmitting only ordinary vibrations. The polaroid P_2 acts as an *analyser* because there is change in intensity for different orientations of P_2.

2. Is it possible to use any other device in place of polaroid?

Ans. Yes, nicol prism.

3. Explain the difference between polaroid and nicol prism?

Ans. Polaroid is a device which polarizes the ordinary light by selective absorption of extra-ordinary vibrations while transmitting the ordinary vibrations. Nicol prism is a device which excludes ordinary vibrations through total internal reflections and transmits extraordinary vibrations.

4. What is the function of a quarter wave plate?

Ans. It is a calcite plate of such a thickness that it introduces a path difference of $\lambda/4$ or a phase difference of $\pi/2$ between ordinary and extraordinary rays when light passes through it. In this experiment this plate is used to obtain the elliptically polarised light.

5. What is a photoelectric cell and how many types of these cells are there?

Ans. A photoelectric cell is a device which converts light energy into electric energy. Known photo cells are : (*i*) photo-emissive cell, (*ii*) photo-voltaic cell (*iii*) photo-conductive cell.

6. Which type of photo-cell have you used in this experiment?

Ans. Photo-voltaic cell. This type of cell does not require any external e.m.f.. This cell is also cheap and quite robust.

7. What is the function of photo-cell in this experiment?

Ans. When tight emerging from the analyser P_2 is allowed to fall on photoelectric cell, a current proportional to the intensity of polarized light starts flowing. Obviously, the intensity of the light can be measured by the deflection of galvanometer.

8. What is the mechanism of emission of photo-electrons?

Ans. When a light photon (energy $h\nu$) is incident on a metal surface, a part of its energy ϕ_0 as work function is used in releasing the electrons from metal surface and the rest energy is utilised in imparting kinetic energy $\left(\frac{1}{2}mv^2\right)$ to the ejected electron. Thus

$$h\nu = \phi_o + \frac{1}{2}mv^2$$

which is famous *Einstein's Photoelectric equation.*

9. What is elliptically polarized light?

Ans. In normal or ordinary light ***E*** (electric vector) vibrates in all possible directions in a plane perpendicular to the direction of propagation. When two plane polarized waves superimposed, the resultant vector of these two light waves rotates with a constant magnitude in a plane perpendicular to the direction of propagation. If the resultant (light) vector varies periodically its rotation such that the vector traces an ellipse, the light is said to be elliptically polarized.

10. What is the order of photoelectric current?

Ans. Few micro amperes.

11. How will you measure current of micro amperes order?

Ans. One can measure such small currents either with the help of a microameter or more accurately with the help of a ballistic galvanometer.

12. What is the order of sensitivity of a ballistic galvanometer?

Ans. $\sim 10^{-9}$ amp.

13. What type of curve one obtains by plotting a graph between intensity and angle of rotation of analyser?

Ans. Polar graph (Fig. 17)

14. Is it possible to read the intensity directly from the galvanometer?

Ans. Yes, provided the galvanometer is standardised for intensity.

EXPERIMENT 9

Object

To determine the height of a object with the help of a sextant.

Sextant

This is an instrument which enables us to determine the angle subtended by two distant point objects at the place of observations. If the distance of one of the points from the sextant is known the distance between the two points can be calculated.

It consists of a graduated circular arm *AB* supported by two fixed radial arms *AC* and *BC* which meet at a point *C* which is the centre of the circle of which the arm forms a part.

An *index arm CD* which is free to rotate round an axis passing through *C* carries a plane mirror M_1 called the *index glass* at the end *C* fixed in a plane perpendicular to the plane of the scale. At the other end of this arm at *D* there is a vernier scale *V* which can slide along the main scale. Ordinarily the vernier scale is fixed in position, but its motion can be made free by pressing the lever at *L*. A tangent screw *S* is provided for fine adjustments of the vernier.

A glass plate M_2, called the *horizon glass*, is fixed on the arm *AO* with its *plane perpendicular to the plane of the scale in such a manner that when the index arm is at zero, this is parallel to the mirror* M_1. Lower half of this glass plate is silvered whereas the upper half is transparent. A telescope *T* is fixed to the arm BC with its axis parallel to the plane of the scale and passing through the centre of the horizon glass. A screw is provided at the side of the telescope (not shown) to move it in a plane perpendicular to the plane of the scale.

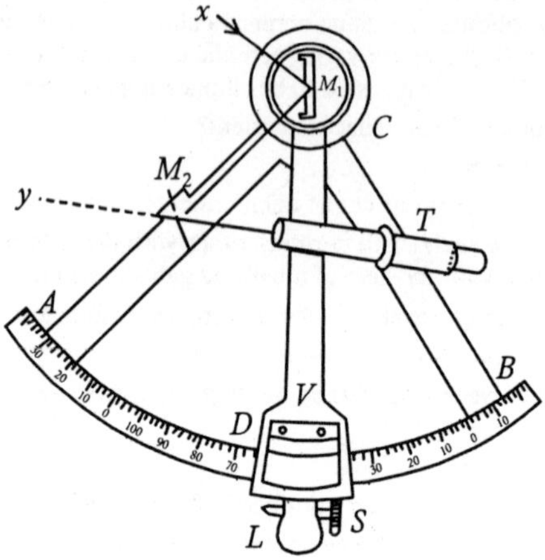

Fig. 19

Working

To find the angular distance between two distant points like x, and y, the reading of the vernier is noted when the image of y as seen direct through the clear portion of M_2 overlaps the image formed after reflection from M_1 and then from the silvered part of M_2. The arm CD is rotated till the image of x as seen after reflection from M_1 and silvered portion of M_2 coincides with the image of y as seen direct. The reading of the vernier scale is again noted. The angular separation between x and y is equal to twice the angle through which the mirror M_1 has been rotated. To facilitate a direct reading from the scale, the arm is graduated in half degrees although markings are in degrees. Such markings are called *sextant degrees.*

Note: In some sextants a micrometer screw is provided in place of the vernier. In such a case in general the smallest division on the scale is *one degree.* The number of divisions on the circular scale of the micrometer is 60. Thus the least count of the micrometer.

$$= \frac{10}{60} = 1'$$

The micrometer is further provided with a vernier having 5 divisions which coincide with 4 divisions of the micrometer scale. Thus the least count using the vernier comes out to be

$$\left(1 - \frac{4}{5}\right) = \frac{1}{5} \text{of a minute} = 12''$$

To Find the Vertical Distance Between Two Points: If h is the vertical distance between two points x and y and θ is the angle subtended by xy at the index glass of the sextant at a distance d as shown in Fig. 20(a), then

$$\frac{h}{d} = \tan\theta$$

or
$$h = d\tan\theta$$

This is true only when $M_1 y$ is perpendicular to xy. From Fig. 20(b).

and
$$\cot\theta_2 = \frac{yQ}{h}$$

$$\therefore \cot\theta_2 - \cot\theta_1 = \frac{yQ - yP}{h}$$

$$= \frac{d}{h}$$

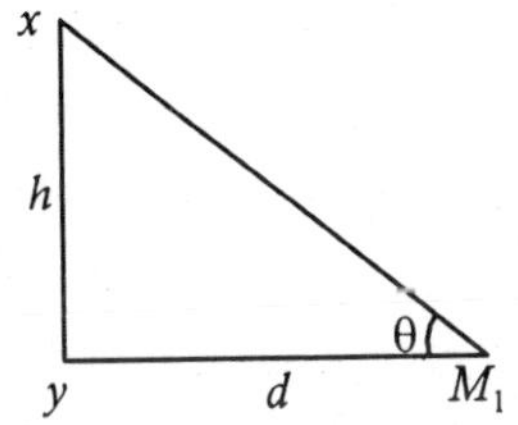

Fig. 20 (a)

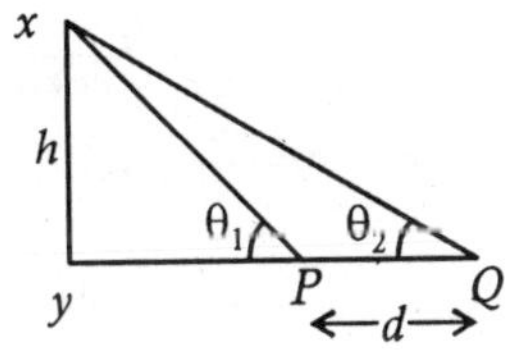

Fig. 20 (b)

Experiment

To find the height of an inaccessible object.

Apparatus

Sextant, Chalk of two different colours, a rigid clamp stand, measuring tape, spirit lamp.

Formula

$$h = \frac{d}{\cos\theta_2 - \cos\theta_1}$$

Procedure

Mark an arrow-head in the same horizontal line as the telescope, at a large distance. This serves as the lower reference point for the inaccessible object. The height of the top of this inaccessible object is determined with respect to this point.

Set the sextant so that the direct and the reflected images of the arrow coincide. Note the reading. When the reflected image of the top of the object coincides with the direct image of the arrow. Note the reading again. Repeat by placing the sextant in three different positions, P, Q and R

in the same straight line by moving the sextant from 5-10 ft. Find the value of h from any pair of observations and calculate the mean value.

Record

Vernier constant = ...

Sextant at	*Scale reading*		*Difference θ*	*cot θ*
	Initial	*Final*		
P Q R			$\theta_1 =$ $\theta_2 =$ $\theta_3 =$	

(*i*) Distance PQ $d_1 =$ cm

$$\therefore \quad h = \frac{d_1}{\cot\theta_2 - \cot\theta_1}$$

(*ii*) Distance QR $d_2 =$ cm

$$\therefore \quad h = \frac{d_2}{\cot\theta_2 - \cot\theta_1}$$

(*iii*) Distance PR $d_3 =$ cm

$$\therefore \quad h = \frac{d_3}{\cot\theta_3 - \cot\theta_1}$$

Mean height $h =$ ...cm.

Precautions

1. Telescope should be at the same height from the floor as is the bottom of the object.
2. The scale should be vertical and the axis of the telescope should be horizontal.
3. The two images should overlap and should have the same intensity.

Questions with Answers

1. What is a sextant?

Ans. This is an instrument which enables one to determine the angle subtended by the two distant point objects at the place of observation. Its scale is only 1/6th of the circle, i.e. its scale forms on arc of 60°.

2. What is the principle of action of the sextant?

Ans. The action of a sextant is based on the principle of rotation of a plane mirror. According to this principle when a plane mirror is rotated through an angle θ about an axis perpendicular to the plane of incidence, the reflected ray rotates through an angle 2θ, the direction of the incident ray remain unchanged.

3. In Question 2, is the incident ray fixed as mirror rotates?

Ans. No, the incident ray is not fixed but the reflected ray is fixed. We may note that on account of the reversibility of light path, the incident and reflected rays are interchangeable. Sextant works on this principle.

4. What are the mechanical conditions, which these two mirrors should satisfy?

Ans. Two mirrors should satisfy : (*i*) Their planes should be perpendicular to the plane of the arc, and (*ii*) When there is no zero error in the instrument, these glasses should be parallel.

5. Explain the functions of the screws attached at the back of the frames carrying mirrors M_1 and M_2.

Ans. These mirrors are made normal to the bed of the instrument with the help of these screws.

6. Explain the function of light filters in sextant.

Ans. Light filters are used to cut down the intensity of light in case the object happens to be too dazzling.

7. What is the total angle which can be read with sextant?

Ans. Upto 120°

8. What type of telescope is fitted with sextant?

Ans. Galilean type of telescope which produces an erect image due to concave lens fitted in its eye piece.

9. Is the actual angle subtended by the arc appears to be an actue angle?

Ans. Yes.

10. How this anomaly explained?

Ans. The values of the angle recorded on the scale are doubled and thus 10° is written as 20°. We know that the angle subtended by the object is twice the angle of rotation of the index glass. Clearly to record the angular elevation, therefore, one will have to double the observed value. In order to avoid this confusion and complexity, the manufactureres have marked the values after doubling them, i.e. the instrument is a direct reading instrument.

11. When a sextant is pointed towards an object, why two images are seen?

Ans. One image is formed due to the rays directly entering the telescope through the transparent portion of the horizon glass,. whereas the second image is formed by those rays which enter the telescope after the reflections from the index glass and the horizon glass.

12. What is zero error in a sextant?

Ans. When the glasses are exactly parallel, the zero of the vernier should coincide with the zero of the main scale. If it is not so, the sextant is said to have a zero error.

13. How will you determine zero error in a sextant?

Ans. One can determine zero error by pointing the sextant towards a distant object, and by coinciding the direct image with the image formed by double reflection at the two glasses. We may note that if the reading of the instrument is zero, then there is no zero error. In case if reading is not zero and this is the zero error.

14. Does the zero error in a sextant depend on the distance of the object?

Ans. Yes, nearer the object, greater is the divergent beam received from it and one finds that greater is the value of zero error. Clearly, zero error is inversely proportional to the distance of the object.

15. What is the use of coloured glasses provided with the sextant?

Ans. We know that the sun or any other brilliant object gives intense glare in the telescope. To diminish this high intensity of light entering the telescope, the coloured glass have been worn in the path of light rays.

16. How you will chose the two points of observation?

Ans. The points of observation be collinear with the reference work.

17. Mention few practical uses of sextant?

Ans. This is used by *navigators* for measuring the angular attitudes of the sun and stars and thus they able to locate their exact positions on the earth. One also uses it to determine the vertical and horizontal distances between terrestrial objects.

EXPERIMENT 10

Object

To determine the transmission coefficient of given transmitting plate using a photometer.

Apparatus

Transmitting plate, Lummer -Brodhum or Bunsen Photometer, Two light sources.

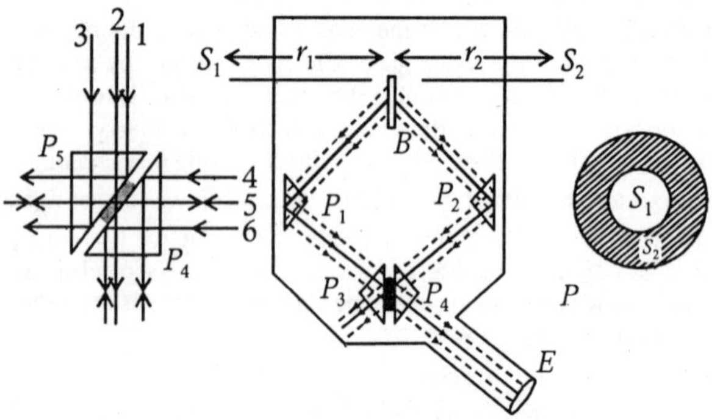

Fig. 21: Lummer-Brodhum Photometer

Description of the Apparatus

It consists of a magnesium carbonate slab *AB* arranged such that each face is illuminated by the sources S_1 and S_2. P_1 and P_2 are the two right angled prisms which receive the light after the reflection from the slab *AB*. The light now reflected from these prisms, enter into a double prism P_3 and P_4. The two prisms are placed with their hypotenuse faces together and separated by a thin air film except at the centre. The central portion is cemented with canada balsam. The rays striking the cemented part are totally transmitted while those striking the air film are totally reflected. This is shown in the left adjoining figure. The ray 1 and 3 coming from prism P_1 are reflected while ray 2 is transmitted. Similarly rays 4 and 6 coming from the prism P_2 are reflected while ray 5 is transmitted. In this way light reflected from prism P_1 are transmitted through P_3 illuminate the central protion of the field of view and light reflected from P_2 and again reflected from P_4 illuminates the rest of the field of view of the telescope as shown in the right in the Fig. 21. Due to the different illuminating powers of the two sources, the intensities of the two portions are different. When the distances of the sources are so adjusted that the two parts are equally bright,

then $$\frac{P_1}{r_1^2} = \frac{P_2}{r_2^2}$$

Theory and Method

Whenever light falls on a surface it undergoes the phenomenon of (*i*) reflection (*ii*) absorption and (*iii*) transmission, i.e. some part of light is reflected, some part is absorbed and some part is transmitted, and if the respective coefficients are, *r*, *a* and *t* respectively, then

$$r + a + t = 1 \tag{1}$$

One can measure the transmission coefficient (t) with the help of Lummer-Brodhum or Bunsen Photometer.

In the first instances the illuminating power of two sources S_1 and S_2 is compared by adjusting their respective distances from the screen S as shown in Fig. 22.

Let the respective distances of the photometric balance be r_1 and r_2. Thus

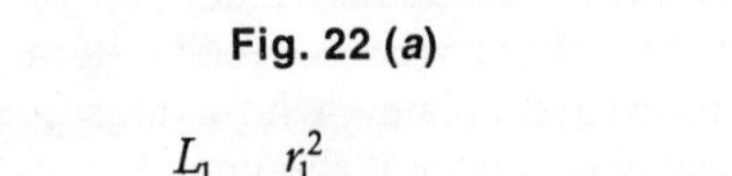

Fig. 22 (*a*)

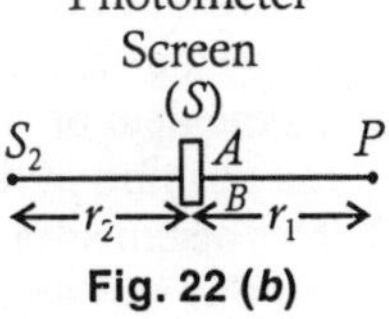

Fig. 22 (*b*)

$$\frac{L_1}{L_2} = \frac{r_1^2}{r_2^2} \tag{2}$$

Now the given transparent plate P whose transmission coefficient is to be determined is introduced on one side of the screen as shown in Fig. 23 and the distance of the source S_2 is again adjusted for photometric balance.

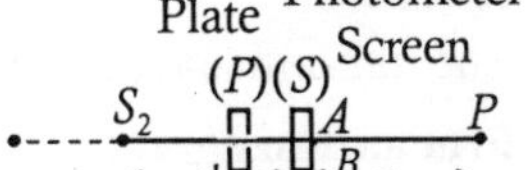

Fig. 23

Let the new position of source S_2 from screen by r_2' and transmission coefficient of plate P be t. In this situation the light transmitted through plate P be considered as a new source having illuminating power L'. So

$$L' = t\,L_2 \tag{3}$$

Under Photometric balance

$$\frac{L'}{(r'_2)^2} = \frac{L_1}{r_1^2}$$

Using Eq. (3) and substituting for L', we have

$$\frac{tL_2}{(r'_2)^2} = \frac{L_1}{r_1^2}$$

or

$$t = \frac{L_1}{L_2}\left(\frac{r'_2}{r'_1}\right)^2$$

Using Eq. (2), we obtain

$$t = \frac{r_1^2}{r_2^2}\left(\frac{r'_2}{r_1}\right)^2 = \left(\frac{r'_2}{r_1}\right)^2 \tag{4}$$

Using Eq. (4), one can determine the transmission coefficient of a transparent surface.

Procedure

(*i*) The photometer is mounted upright nearly in the middle of an optical bench and adjusted in such a way that photometer head becomes normal to the line joining the two sources as shown in Fig. 22(a).

(*ii*) Remove the shutters from both sides of photometer and allow the light to fall on each face of the head.

(*iii*) One of the lamp is fixed and the position of this lamp and photometer head are noted. For one set of observation this should not be disturbed.

(*iv*) Move the other lamp from one end of the bench towards photometer till the field of view in telescope is equally illuminated. Note down the position of the lamp.

(*v*) The distance of the first lamp is changed by 5 cm and the other lamp is adjusted till again the field of view is equally illuminated.

(*vi*) Now introduce the given transparent plate P whose transmission is introduced on the side of the photometer as shown in Fig. 24. Now adjust the distance of the source S_2 for photoelectric balance. Repeat this four times and note down the readings.

Result

The transmission coefficient of a transparent plate is ...

Precautions

(*i*) Heights of the two sources and photometer should be carefully adjusted.

OBSERVATION TABLE

S.No.	*Distance of source S_1 from screen S for photometric balance (r_1)*	*Distance of plate P from screen S for Photometric balance (r_2')*	$t = \left(\frac{r'_2}{r_1}\right)^2$
1			
2			
3			

Mean t =

Questions with Answers

1. What is illuminating power ?

Ans. The illuminatig power of a source is the quantity of light incident per second on a unit area placed at a unit distance from the source in a direction normal to the rays. It is measured in candle power.

2. What is intensity of illumination ?

Ans. The intensity of illumination at a point is defined as the light incident per second on a unit area of the surface placed at the point under consideration.

$$\text{Intensity of Illumination} = \frac{\text{Candle Power}}{(\text{Distance})^2}$$

3. How are illuminating power (P) and ; the intensity of illumination (I) are related?

Ans. $I = P/r^2$

where r is the distance of the point from the source.

4. Is intensity of illumination at a point follows inverse square law?

Ans. Yes, $I \propto 1/r^2$, where r is the distance of the point from the source

5. Is human eye is equally sensitive to all colours?

Ans. No, in the visible range the human eye is not equally sensitive to all colours Figure 24 shows the variation of sensitivity with wavelength. We may note that it is maximum at 5550 Å in the yellow region.

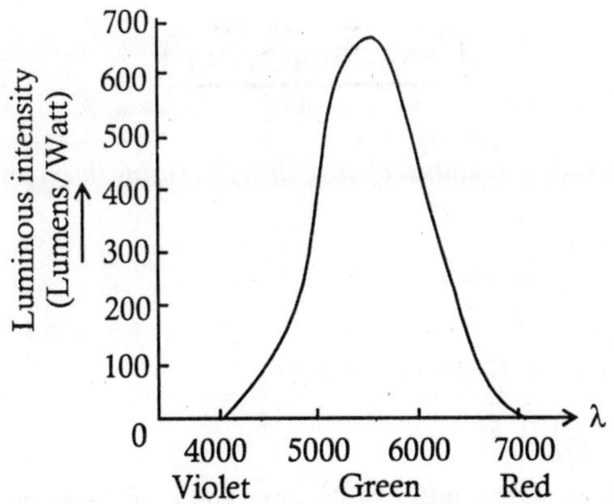

Fig. 24: Variation of luminious intensity with wavelength (λ)

6. Can our eye distinguish two surfaces placed side by side and seen simultaneously?

Ans. Yes, if their brightness differ by 1 %.

7. Can you use Lummer Brodhum photometer for the comparison of illuminating power of two sources emitting light of different colours?

Ans. No, human eye cannot judge the equality of contrast or brightness in two different colours.

8. Out of the two versions, viz. in 'equally brightness type' or contrast type of Lummer Brodhum photometer, which is better and why?

Ans. The contrast type photometer is better because the eye can judge the uniformity of contrast more accurately than the equality of brightness.

9. Which substance is used for the screen of photometer?

Ans. Magnesium carbonate.

10. Why is the colour of the diffusing screen white?

Ans. A white surface reflects equally well and hence the light diffused by it does not differ appreciably from the original.

11. What is the purpose of repeating the observation after rotating the diffusing screen through 180° in a Lummer Brodhum photometer?

Ans. In order to eliminate error due to difference in the reflecting powers of the two surfaces of the diffusing screen.

12. On what factors does the intensity of illumination (I) depend?

Ans. $I = P\cos\theta/r^2$

Here $P \to$ luminous intensity, $\theta \to$ angle that normal to the surface makes with the direction of r and $r \to$ the distance of point under consideration. Obviously I is (*i*) directly proportional to P (*ii*) inversely proportional to r and (*iii*) directly proportional to $\cos\theta$, this is known as *Lambert's cosine law.*

13. Define transmission coefficient of a transmitting plate.

Ans. The ratio of the illuminating power of a light source in the presence of given transmitting plate to the illuminating power of the same source in the absence of the said plate for uniform illumination is called the transmission coefficient.

EXPERIMENT 11

Object

To measure numerical aperatuse of an optical fiber using fiber optic training kit.

Apparatus

Optical fibre training kit, Screen.

Formula

Numerical aperatuse (NA) is given by:

$$NA = \left(n_1^2 - n_2^2\right)^{1/2}$$

where n_1 is the refractive index of the core and n_2 is the refractive index of the cladding.

Procedure

When a cone of light is incident on one end of the fibre, it will be guided through it, provided the semi-angle of the cone is less than in (the maximum value of sin i for ray guided is given by $\sin i_m = \sqrt{n_1^2 - n_2^2}$ when $n_1^2 > (n_2^2 + 1)$ and $\sin i_m = 1$ (when $n_1^2 > (n_2^2 + 1)$.

This angle is a measure of the light gathering power of the fibre. For

$$n_1 = 1.48,\ n_2 = 1.46,\ NA \approx 0.24, \Rightarrow i_m \approx 14°$$

EXPERIMENT 12

Object

To study basic properties of laser and Holography using Laser Source.

Apparatus

Green Helium-Neon Laser, Screen.

Procedure

Laser gives a very highly intense and highly collimated monochromatic beam of light. A small amount of smoke made by burning some dhoop stick in its path is sufficient to make visible a ray of light from the laser passing across the room. Two laser sources can be used to study the holographic properties.

If we put a prism or a glass slab in the path of the laser beam produces diffraction pattern on the screen held at a distance infront of the slit. The pattern shows central maximum intensity spot and secondary maxima of decreasing intensity separated by dark spaces.

When a diffraction grating is placed in the path of the laser beams, a central spot is seen on the screen and on the two sides of this spot there are three orders of decreasing intensity spots.

Note:

1. A green helium-neon laser is available in the market for about Rs. 6000/-.
2. All those experiments which we do with ray box can be done with the laser beam. The advantage here is that with a small amount of smoke the rays are clearly visible passive across the room.

Precautions

1. One should avoid the direct looking into the laser beam or putting skin in the direct path of laser beam.

2. It is better to keep the laser source and optical instruments on the floor and observe them from taking refracted or diffracted rays on the screen.

Experiment I

Demonstration of the spatial coherence of the laser beam.

Equipments

A gas laser, a variable double slit (or a number of double slits with different separations), a viewing screen.

Procedure

The experiment is usually done with the unexpended beam that is allowed to travel sufficiently large distance so that it is of reasonable size (say about 5 mm in diameter) for the slits to insert and move. Alternatively it may be lightly expanded. The spatial coherence in the beam can be demonstrated by any one of the following procedures.

The variable double slit is inserted within the laser beam and the interference pattern is observed on a screen placed several meters away from the slit plane. Young's fringes are straight line fringes with spacing of

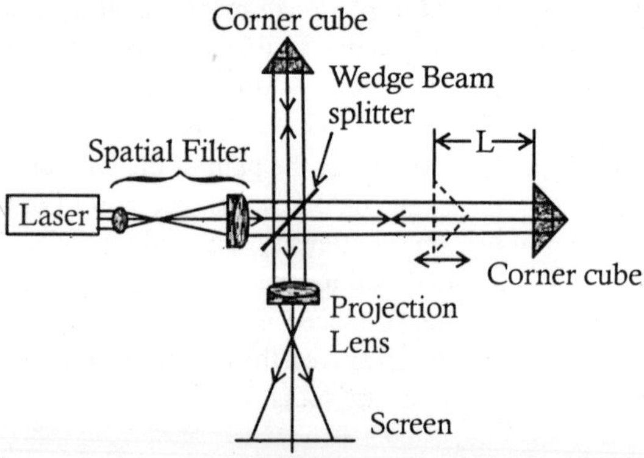

Fig. 25: Demonstration of Temporal Coherence

$\bar{x}$ ($= \lambda z/d$), where z is the separation between the slit plane and screen and d is the slit separation. The fringes run parallel to the slits. The central fringe in the pattern is bright. The fringe pattern can be scanned with a photodetector carrying a tiny pin-hole. The visibility is calculated by measuring the irradiance of the central bright fringe and the minima around this. The visibility is calculated from the formula $V = (I_{max} - I_{min})/(I_{max} + I_{min})$. The visibility is calculated around this. The visibility is measured as a function of slit separation. If the laser is oscillating in TEM_{00} mode, the visibility remains practically constant for all slit separations spanning the beam. If the laser is oscillating in TEM_{10} mode, the interference pattern remains stationary so long the slit pair is within one of the lobes of the beam. When the slits lie on both the lobes, the central fringe is dark; there is a shift of half fringe width when the slit pair crosses from a single lobe over to both the lobes.

The spatial coherence can also be demonstrated by scanning the laser beam with a double slit of fixed separation. It will be observed that the fringe pattern will remain stationary regardless of which portion of the beam is intercepted as long as beam fills the double slit. This demonstrates that there is always fixed phase relationship between the two waves from the double slit. The laser beam is thus spatially coherent across the entire beam.

Experiment II

Demonstration of the temporal coherence of the laser beam.

Equipments

A gas laser, an interferometric set up (Michelson interferometer), a screen, etc.

Procedure

It is assumed that the laser is oscillating in TEM_{00} mode so that spatial coherence exists over the whole beam. The interferometric set up (Fig. 25) consists of a microscope objective, a collimating lens, a beam splitter preferably wedge type, two plane mirrors (or cube corners) and a projection lens. The beam from the laser is expanded to about 25 mm in diameter using microscope objective and collimating lens combination. The collimation can be checked. The collimated beam is split into two beams of approximately equal irradiances by the beam splitter. The beams proceed towards the two plane mirrors one of which is mounted on a translatory stage. The reflected beam recombine at the beam spliter. If we place a screen at the focal plane of the projection lens, two bright spots corresponding to the two reflected beams are seen. The two spots are made coincident by adjusting the mirrors on the their adjustable mounts. The interference pattern is now visible on the screen which is now placed away from the focal plane of the projection lens. The pattern contains straight line fringes whose orientation and number can be controlled by adjusting the plane mirrors.

Initially both the mirrors are kept at approximately equal distance from the beam splitting surface of the beam splitter. This corresponds to zero time delay between the two beams. The fringes of maximum contrast are, therefore, seen. The mirror on the translatory stage is now moved away parallel to itself and the visibility of the fringes as a function of path difference is studied. If the laser is oscillating in a single longitudinal mode, their is no apparent decrease in the visibility over very large distances, demonstrating a very large temporal coherence. If the laser oscillates into two longitudinal modes, the visibility would exhibit minima at the path difference which is multiple of the cavity length.

Often the exercise is done by a student prior to setting up an holographic recording arrangement. The contrast of interference pattern is observed by setting up Michelson interferometer with unexpanded beam and moving one of the mirrors in discrets steps. The procedure described above using expanded beam and continuous motion of the mirror is obviously more accurate.

In this experiment, it should be noted that the moving mirror may tilt during traverse due to imperfect guides, and hence the interference pattern may disappear. The traverse guides on which the mirror moves should be extremely accurate. If the cube comer is substituted for the plane mirror, the interferometer becomes tilt insensitive. The long coherence length of the laser beam with cube corners as reflectors in interferometric set up has been fully exploited in numerically controlled machines.

Experiment III

Recording of a hologram and reconstruction of the image.

Equipments

A gas laser (1 to 10 mW), two microscope objectives, one beam-splitter, two front coated plane mirrors, a plate/film holder, holographic plate/ film and processing facilities.

Procedure

There are many ways to record a hologram. We shall, however, describe a procedure which is followed by us. Various optical components are mounted on the holographic table as shown in Fig. 26. It is not necessary to mount the laser on this table, only the components from the beam-spiltter onwards should be on the vibration isolation table. Care should be taken to match the path lengths *ABCD* and *AED*. The beam-spiltter divides the beam into two parts; the beam 1 is used as reference and the beam 2 is used to illuminate the object as shown in Fig. 26. The beam 2 should be stronger than the beam 1 as the scattered field from the object is usually very weak. The reflected beam from the second surface of the beam spiltter which also goes in the direction of beam 1 can be eliminated by inserting a stop in the path. Use of wedge type beam spiltter would be convenient. The mirrors M_1 and M_2 are used to fold the paths. It may be noted that the beam splitter, mirrors M_1 and M_2 need not be of good optical quality as the laser beam of approximately 1 mm diameter is incident on them. A pin-hole and microscope objective combination, often called a spatial filter, is inserted in both the beams. The spatial filter removes the high frequency contents in the beam which could arise due to diffraction from dust particles with the optical elements and the spurious interference fringes. The pin-hole of approximately 20 μm diameter is used. Since the microscope objective focuses the beam at pin-hole, a very accurate alignment of pin-hole and microscope objective is required. The pin-hole is provided with a *x-y* motion and the microscope objective with *z*-motion. The beam 1 emerging from the spatial filter constitutes a spherical reference wave. The beam 2 from the spatial filter illuminates the object.

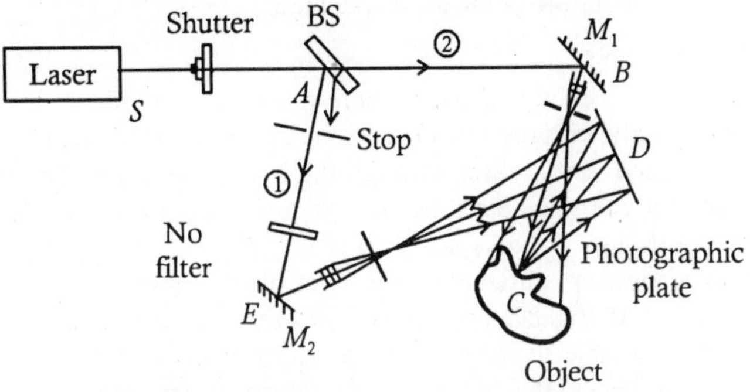

Fig. 26: Schematic Reading of a Hologram

The experiments is carried out in dark. The laser beam is suitably blocked and the photographic plate is inserted in the plate holder which is rigidly mounted on the table. The plate cover is now removed and a few minutes are allowed for the disturbance to subside before the exposure is made by letting the beam remotely. With a 5mW laser and a diffuse object 20 cm away from the plate, an exposure time of the order of a few seconds is required. If the powers of the beams at the plane of the photographic plate are measured, the exposure time is calculated from the technical data of the plate emulsion supplied by the manufacturer. After the proper exposure has been given, the laser beam is blocked and the plate is removed to the dark room for processing.

Processing

Following procedure yields satisfactory results for Kodak and Agfa Gevacrt emulsions.

Development

5 min in Kodak *D*-19 developer with continuous agitation. Follow immediately with 30 sec rinse is water.

Fixing

5 min in Kodak rapid fixer bath with continuous agitation. Follow immediately with 30 sec rinse in water.

Residual Fixer Removal

1.5 min in Kodak hypo clearing agent.

Wash

5 min in running water.

Dry

Drying in clean warm air.

The recommended temperature of the developer bath is 20° C ± 0.5° C. The temperatures of other chemical baths are not critical but are usually kept at the temperature of the developer bath.

Reconstruction

The hologram is mounted back where it was for recording. It is now illuminated only with the reference wave, i.e. the object is removed and beam 2 blocked. On looking through the hologram in the direction of the object one sees the object, as through a window exactly at the same location, although it has since long been removed. By changing the direction of viewing, different perspectives of the image can be seen and photographed. In Fig. 27. a photograph of a reconstructed image is given. It may be advisable to reconstruct the image using a spectral lamp as accidental viewing in the direction of the reference wave may be injurious to the eye when laser light is used for construction.

Fig. 27: Photograph of a Reconstructed Image

EXPERIMENT 13

Experiment

To measure the focal length of the combination of two lenses separated by a distance with the help of nodal slide.

Apparatus

Nodal slide, source of light, optical bench, metallic cross slit, two convex lenses and a plane mirror.

Formula

$$\frac{1}{F} = \frac{1}{f_1} + \frac{1}{f_2} - \frac{d}{f_1 f_2}$$

where F is the focal length of the combination of lenses, f_1 and f_2 are the focal lengths of the two given lenses and d is the separation of the two lenses.

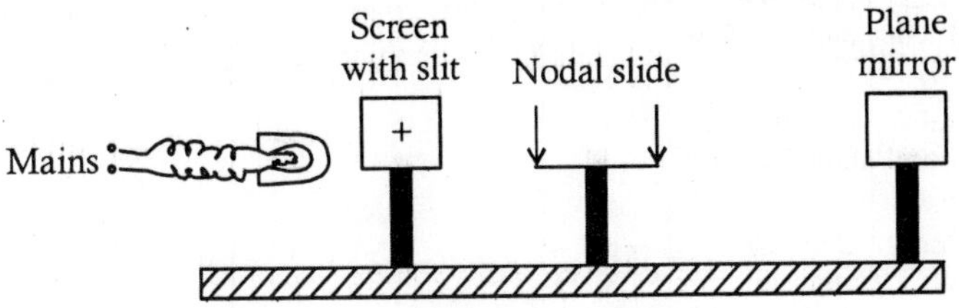

Nodal Slide Assembly

Fig. 28

Method

First adjust the lamp, cross slit and mirror in such a manner that their axis lie along the same horizontal line. We must keep the lenses at a known separation, say d. Now adjust the nodal slide such that both the lenses are at equidistance from the centre of stand. This helps to measure the distance of lenses easily on the bench scale. Now, we move the entire lens system as a whole towards or away from the slit till a clear and proper image of cross slit is formed on the same screen of slit. The distance between slit and centre point of combination of lens in this position is the focal length (F) of combination of lenses. One can judge the exact position by rotating the nodal carriage about 5° from its position, the image of cross slit should not move. Now, repeat the above procedure after rotating the nodal carriage. The procedure is also repeated with changing the distance between the lenses. One can also find the focal length of single lens, i.e. f_1 and f_2 in similar way.

Calculations

The focal length of the combination of lenses is given by

$$\frac{1}{F} = \frac{1}{f_1} + \frac{1}{f_2} - \frac{d}{f_1 f_2}$$

or $F = \ldots$cm.

Result and Conclusion

The calculated value of the focal length of the combination = ...cm.
The experimental value of the focal length of the combination = ...cm.
The two values are approximately same.

Observations

TABLE 1

Lens	*Light incident on face*	*Distance between two lenses* (d) cm	*Position of cross slit* (a) cm	*Position of Nodal stand* (b) cm	*Focal length of the lens* ($a-b$) cm	*Average focal length of the lens* (cm)
First lens (focal length f_1)	One face opposite face					
Second lens (focal lenght f_2)	One face Opposite face					
Combination of lenses	One face opposite face					

Sources of Errors and Precautions

(*i*) Care must be taken while focusing the image of the cross slit.

(*ii*) Care must be taken in reading the positions of the mark on the lens holder.

Questions with Answers

1. What is a nodal slide?

Ans. It is a small horizontal metal carriage having two lens holders. The entire arrangement as a whole can be moved back and forth in horizontal direction as well as can also be rotated about a vertical axis.

2. Why we call it a nodal slide?

Ans. Nodal slide is used to locate the nodal points of a lens system and hence it is called nodal slide.

3. What are cardinal points?

Ans. Two focal points, two nodal points and two principal points together are called cardinal points.

Nodal Points: A pair of conjugate points on the principal axis of coaxial lens system having unit positive angular magnification.

Focal Points: A pair of focal points lying on the principal axis of a coaxial lens system are called focal points.

Principal Points: A pair of conjugate points on the principal axis of the coaxial lens system having unit positive linear or lateral magnification.

4. What do you understand by a coaxial lens system?

Ans. A coaxial lens system is a combination of two or more lenses having common principal axis.

5. Define angular magnification.

Ans. Angular magnification $= \dfrac{\tan\theta_2}{\tan\theta_1}$

where θ_1 and θ_2 are angles made by the incident and conjugate emergent rays with the principal axis.

6. Define lateral magnification.

Ans. Lateral magnification $= \dfrac{\text{length of the image}}{\text{length of the object}}$

7. Explain the importance of the cardinal points of the coaxial optical system.

Ans. Using cardinal points, one can get the size of the image of an object directly by using simple formula developed for thin lenses.

8. What is the condition under which six cardinal points of an optical system reduce to four?

Ans. When the medium on either side of the optical system and also between both the lenses is same, the principal points coincide with the nodal points. Thus the six cardinal points reduce to four.

9. Will the relative positions of nodal and principals change, if the whole system be dipped in water?

Ans. As the medium on either side of the system is the same and hence there may not be any change.

10. What is the function of plane mirror in this experiment?

Ans. Plane mirror reflects the parallel rays incident on its plane.

11. Suppose the position of plane mirror is changed, how this will affect the final image?

Ans. The position of plane mirror does not affect the position of the image.

12. On what factors does the focal length of a thin lens depend?

Ans. Focal length (f) of a thin lens depends upon: (*i*) Refractive index (μ) of the material of the lens (*ii*) radii of curvature R_1 and R_2 of its two surfaces.

13. Is it essential to rotate the lens system by 180°?

Ans. The combination of two thin lenses should be rotated by 180° because the nodal points are not symmetrical with respect to two thin lenses, since they are of different focal lengths.

EXPERIMENT 14

Object

To measure the viscosity of water by capillary tube flow method.

Apparatus

Constant head apparatus, beaker, stop-watch, capillary tube, rubber tubes for connecting purpose, meter scale, spirit level, etc.

Formula

$$\eta = \frac{\pi\rho g r^4}{8l}\left(\frac{h}{Q}\right)$$

Symbols have usual meanings

Method

With the help of rubber tube, connect the capillary tube with the outlet of constant head apparatus. With the help of spirit level, it stands horizontally on wooden support. Now, turn on the water supply so that the water drips at a slow rate from the open end of capillary tube. One can achieve this by adjusting the pinch cock over the rubber tubing so that flow of water through capillary tube is uniform. Now, collect the water in a measuring flask and record the time with the help of stop watch. Also note down the difference of the water level in manometer. By changing the height of constant level water tank, repeat the experiment with the help of microscope, measure the radius of the capillary in vertical as well horizontal directions.

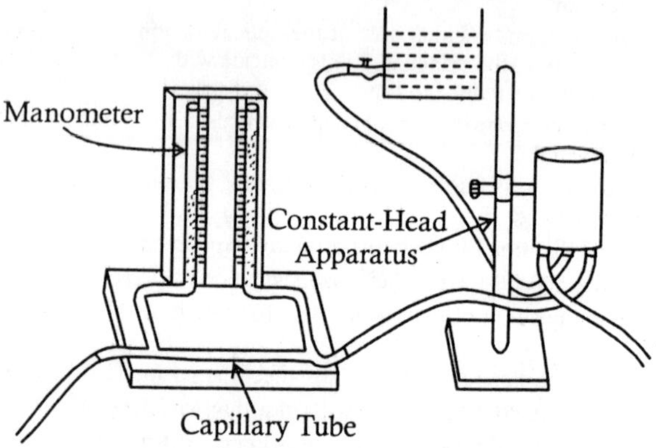

Fig. 29: Capillary tube flow apparatus

Observation and Measurement

Length of the capillary tube = ...cm.
Temperature of water = ... °C

OBSERVATION TABLE 1 FOR *h* AND *Q*

S.No.	h_1 (cm)	h_2 (cm)	$h_1 - h_2 = h$ (cm)	V *Volume of water collected* (cc)	*Time* t (secs)	$Q = \frac{V}{t}$	$\frac{h}{Q}$

OBSERVATION TABLE 2: RADIUS OF CAPILLARY TUBE

S.No.	*At one end of capillary*		*At other end of the capillary*	
	Along any diameter	*Along perpendicular direction*	*Along any diameter*	*Along perpendicular diameter*

Calculations

$$\eta = \frac{\pi \rho g r^4}{8l} \frac{h}{Q} \text{ poise}$$

where ρ and g are density of water and acceleration due to gravity respectively.

$$\eta = \ldots \text{ poise}$$

Result

The coefficient of viscosity of water, η at ...°C temperature is = ... Poise.

Sources of Errors and Precautions

(*i*) Measure h carefully

(*ii*) Measure the radius of capillary carefully, as it appear in the formula with fourth power. A small error in this measurement may affect the result adversely.

(*iii*) Measure the time for collecting water correctly.

Questions with Answers

1. How the constant flow of water is maintained during the experiment?

Ans. With the help of constant level water tank.

2. Why is it necessary to keep the pressure difference constant between the ends of the capillary tube?

Ans. To overcome the viscous force acting between different layers of the liquid, one will have to maintain constant pressure difference

3. Can we connect directly the capillary tube with a water tap?

Ans. No, if we connect the capillary tube directly, the pressure difference across the ends of we tube will be very large and as a result there may be *turbulent* flow through the tube.

4. Why is a long tube chosen?

Ans. For streamline flow.

5. Is this method suitable for the determination of coefficient of viscosity for all liquids?

Ans. This method is only suitable for mobile liquids.

6. What is the aim behind keeping the capillary tube horizontal?

Ans. So that flow of liquid through the capillary may not be affected. If the tube is not horizontal the flow will be affected by the gravity of the earth.

7. Why is the capillary tube of uniform radius used?

Ans. To keep the flow of liquid streamline.

8. If we use capillary tube of large diameter, what will happen?

Ans. Flow of liquid may become turbulent, since the rate of flow of liquid becomes quite large even for a small pressure difference.

9. On what factors does the critical velocity depend?

Ans. $v_c = \frac{Kn}{\rho r}$, where K is Reynold's number, ρ the density of the fluid and r the radius of the tube. We define critical velocity (v_c) as that maximum velocity of a fluid above which a streamline flow changes to a turbulent flow. v_c depends on K, η, ρ and r. For laminar flow K lies between zero and 2000 and for a narrow tube $K = 1000$.

10. What do you understand by a steady state and turbulent flow?

Ans. If each particle in a liquid passing through a certain point follows exactly the same path and has the same velocity as its preceding particle, the flow is said to be steady. We may note that in steady flow, the velocity can be different at different points but at particular point, the velocity is always constant.

In a turbulent flow, the velocity of liquid molecules at a point keep on changing in random way in magnitude as well as direction.

11. What is coefficient of viscosity?

Ans. η is equal to the viscous force acting between the two adjacent layers of unit area having unit velocity gradient between them.

12. What is the unit of η?

Ans. Poise in C.G.S. unit, Ns/m^2 in SI system and one SI unit of viscosity = 10 Poise.

13. Define Poise.

Ans. It is defined as the value of coefficient of viscosity attained by a liquid when a tangential force of 1 dyne per square cm is required to maintain a relative velocity of 1 cm/s between the two layers of the liquid at 1 cm apart.

EXPERIMENT 15

Object

To measure coefficient of viscosity of a liquid by rotating cylinder method.

Apparatus

Searle's rotating cylinder, viscometer apparatus, glycerine (experimental liquid), weight box, stop watch, thermometer and vernier callipers.

Formula

$$\eta = \frac{mgd\left(r_2^2 - r_1^2\right)}{8\pi^2 r_2^2 r_1^2} \frac{T}{l}$$

where symbols have usual meanings.

Method

With the help of vernier callipers, first measure the radii of the inner and outer cylinders r_1, and r_2 respectively and also the diameter of the drum (d). Now pour the experimental liquid (whose viscosity is to be measured) into the outer cylinder. Now, insert the inner solid cylinder in the liquid and tight its ends with the help of lubricated screws to reduce the friction. The drum is attached to the inner cylinder on which the number of turns of the threads carrying the pan S with weight are wound round it. By introducing the stop into the punch hole the whole system is clamped. Now, we count the number of rotation of cylinder and record the time (t) for unwounding thread for a particular lens (l) of the cylinder immersed in the liquid. One can change the length of the inner cylinder dipped in the liquid by raising or lowering the outer cylinder. Repeat these observations for different m and number of turns of thread over the drum.

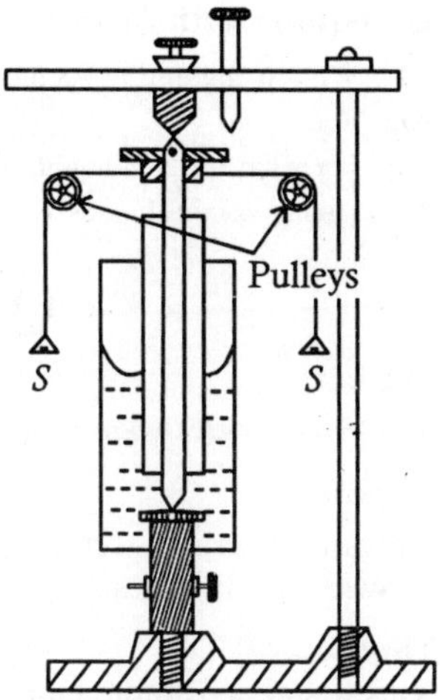

Fig. 30: Rotating Cylinder Method

Observation

Room temperature = ... °C
Inner cylinder radius (r_1) = ... cm
Outer cylinder radius (r_2) = ... cm
Diameter of the drum (d) = ... cm
Mass of each Pan (W_1) = ... gm

TABLE 1

Mass on scale pan W_2 *(gm)*	*Total mass* $m = (W_1 + W_2)$ *(gm)*	*Length of inner cylinder immersed,* l (cm)	*No. of rotations*	*Time taken in* (sec)	*Time period* $T = t/n$

Calculations

$$\eta = \frac{mgd(r_2^2 - r_1^2)}{8\pi^2 r_1^2 r_2^2} \frac{T}{l}$$
$$= \ldots \text{ Poise}$$

Result

The coefficient of viscosity of the given liquid (glycerine) at 0°C is ... Poise.

Sources of Errors and Precuations

(*i*) Care must be taken in measuring the radii of inner and outer cylinders.
(*ii*) Error may occur in measurement of timing when thread just unwounds.
(*iii*) We may note that to get frictionless ends of the inner cylinder is impossible and therefore, there are chances of errors.

Questions with Answers

1. What is the unit of coefficient of viscosity?

Ans. Poise.

2. For what types of liquid, this method is suitable?

Ans. Highly viscous liquids.

EXPERIMENT 16

Object

To convert a galvanometer into an ammeter of a given range and to caliberate it.

Apparatus

Galvanometer, an accumulator, an ammeter (of some range as given for conversion), high resistance box, a low resistance box, screw gauge.

Theory

An ammeter is simply a low resistance galvanometer. Since a galvanometer gives a full scale deflection for a very low current, a shunt (suitable low resistance in parallel) is placed with galvanometer coil so as to measure given range of current with proper caliberation.

The shunt resistance S required to convert galvanometer of resistance G into an ammeter of range I amp is given by

$$S = \frac{I_g}{I - I_g} G \qquad (1)$$

where I_G is the current for full scale deflection of the galvanometer.

The length l of the resistance wire to be used as the shunt with the galvanometer coil is given by

$$l = \frac{S\pi r^2}{\rho} \qquad (2)$$

where r is the radius of the shunt wire and ρ the specific resistance of meterial of shunt.

Method

(A) Determination of Galvanometer Resistance and the Current Sensitivity:

The galvanometer resistance is determined by half deflection method.

1. Connections are made as shown in Fig. 31.
2. Let θ be the galvanometer deflecting when switch S_1 is closed and S_2 open when resistance R is introduced from high resistance box (~ 5000Ω).
3. Now switch S_2 is closed and resistance r from low resistance box (~0 to 50 Ω) is adjusted till the deflection in galvanometer becomes $\theta/2$. The galvanometer resistance $G = r$.

4. Calculate the current sensitivity for the galvanometer and hence I_g the current for all scale deflection.

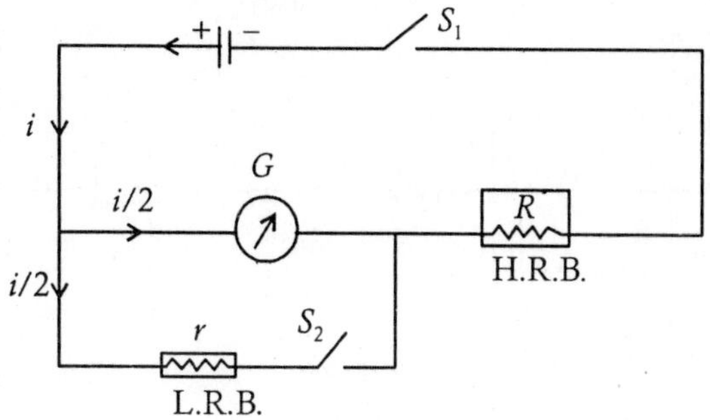

Fig. 31

$$\text{Current sensitivity} = \frac{\text{Deflection (in divisions } n)}{\text{Current } I}$$

$$\text{and } I_g = \frac{\text{Current } I}{\text{Deflection (in divisions } n)} \times \text{Total no. of div. on gal. scale } (N)$$

$$= \frac{E}{R+G} \cdot \frac{N}{n} \qquad (3)$$

where E is the emf of the accumulator.

Method

(B) Caliberation of Shunted galvanometer

1. Calculate the shunt resistance S using eq. (1).
2. Calculate the length of the shunt wire required for converting galvanometer into ammeter using eq. (2).
3. Connect this shunt wire to the galvanometer in parallel.
4. Make the connections as shown in Fig. 32 in which box shows the shunted galvanometer.
5. Adjust resistance and close the switch 'S' as down in Fig. 32. Note the readings of ammeter as well as galvanometer.
6. By adjusting the resistance R, the full scale of galvanometer is caliberated.
7. Find the difference of the ammeter reading and galvanometer reading.
8. Draw a caliberation curve for the galvanometer by drawing graph between ammeter reading and corresponding galvanometer reading. (Fig. 33).

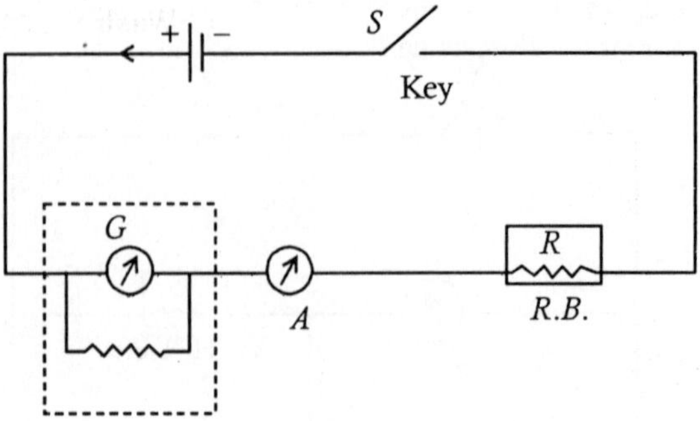

Fig. 32

Observations

Range of Conversion = ... amps.
emf of the cell E = ... Volts,.
Total number of divisions of galvanometer scale N = ... Div.
Specific resistance ρ of the meterial of the wire used = ... Ohm-meter.

(A) DETERMINATION OF G-GALVANOMETER RESISTANCE AND I_g.

S. No.	*Resistance R (Ohms)*	*Galvanometer deflection n (Div.)*	*Resistance r for the deflection (Ohms)*	*Galvanometer resistance G = r (Ohms)*	$I_g = \frac{E}{R+G} \cdot \frac{N}{n}$ *(amp)*

Mean I_g = ...amps.

(B) DETERMINATION OF RADIUS OF WIRE

Least count of the screw gauge = ...
Zero correction = ...

S. No.	*Readings in one direction (Diameter)*				*Reading in mutually perpendicular directions (Diameter)*				*Diameter*
	M.S.R.	*C.S.R.*	*Corrected C.S.R.*	*Total (a)*	*M.S.R.*	*C.S.R.*	*Corrected C.S.R.*	*Total (b)*	$\frac{a+b}{2}$

Mean diameter = ... Mean radius = ...

(C) CALIBRATION OF SHUNTED GALVANOMETER

After conversion one division of galvanometer scale ... amps.

S. No.	*Ammeter Reading I (amp)*	*Galvanometer reading*		*(I' –I) amp (error)*
		Div.	*I' Amperes*	

Calculations

1. $I_g = \frac{E}{R+G} \cdot \frac{N}{n}$ amp
2. $S = \frac{I_g}{I - I_g} G$ Ohms
3. $l = \frac{\pi r^2 S}{\rho} m$

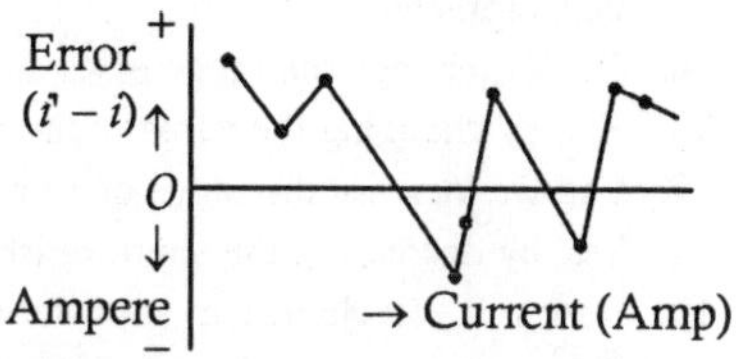

Fig. 33: Ammeter calibration curve

Results

1. Shunt resistance S to be used with galvanometer = ... Ohms
2. Length of shunt wire = ... m
3. Caliberation curve is plotted. You may note that caliberation curve for different ammeters may be different.

Precautions

1. The accumulator used should be fully charged so that its emf remains constant throughout the experiment.
2. Proper care should be taken in connecting the shunt, so that an exact length is connected in between the galvanometer terminals.
3. At no time the resistance from high resistance box be reduced to zero, otherwise in galvanometer, high current will flow and it would be damaged.

Questions with Answers

1. What is ammeter?

Ans. This is an electrical instrument designed to measure current in an electric circuit.

2. How ammeter is connected in a circuit?

Ans. In series.

3. If ammeter is connected in parallel to an electric circuit, what will happen?

Ans. The ammeter will only measure a part of the current flowing through the circuit and not entire current.

4. What is the resistance of an ideal ammeter?

Ans. Zero. However, in practice it is not possible to achieve an ammeter with zero resistance. It is made as low as possible.

5. Can we use galvanometer in place of an ammeter in a circuit?

Ans. No. Due to following reasons it is not possible.

(*i*) The resistance of the galvanometer is somewhat finite. This is when galvanometer is connected in series, the circuit resistance increases thereby decreasing the value of the current to be measured.

(*ii*) Galvanometer being highly sensitive instrument, can measure currents of the order of only microamperes.

6. What is the order of resistance of a moving coil ammeter?

Ans. $10^{-2}\,\Omega$.

7. How can a galvanometer be converted into an ammeter?

Ans. For a given range, one must find the resistance of the shunt. After knowing the resistance, the length of the shunt is calculated. This is put in parallel to the galvanometer.

8. Can we change the range of an ammeter?

Ans. Yes, by changing the value of the shunt resistance.

9. Can we increase the range of an ammeter?

Ans. Yes, by decreasing the shunt resistance.

10. How will you change an ammeter into milliammeter?

Ans. By increasing the value of shunt resistance. An milliammeter is a low range ammeter.

11. By what method have you determined the resistance of the galvanometer?

Ans. Half deflection method.

12. On what principle does moving coil galvanometer work?

Ans. If a current carrying coil is suspended in a magnetic field, a torque acts on it under the influence of which the coil gets deflected. The deflection produced in the coil is directly proportional to the current flowing in it. This is the principle.

13. Other than half deflection method, is there any other method for the determination of galvanometer resistance?

Ans. Yes, with the help of post office box.

13. How many types of galvanometers do you know?

Ans. (*i*) Moving suspended coil type

(*ii*) Moving pivoted coil type

14. Do you know any other more accurate method for caliberating the converted ammeter?

Ans. Yes, with the help of a potentiometer.

15. Which type of galvanometer is generally used in laboratory?

Ans. Pivoted moving coil weston galvanometer.

16. How many types of suspended coil galvanometer are there?

Ans. (*i*) Ballistic galvanometer and (*ii*) Dead-beat type galvanometer.

17. What is the difference between ballistic and dead beat galvanometer?

Ans. In ballistic, the coil is wound on non-conducting core. When a current pulse is allowed to flow throuh it, the coil starts oscillating and since there are no eddy currents, the oscillations are maintained.

In dead-beat, the coil is wound on a metallic core, so that the eddy current produced in the core retards the motion of the coil.

18. In any circuit, does the main current flows through the galvanometer coil?

Ans. No, only a fraction of very small current flows through it.

19. What are the resistance of dead-beat and ballistic galvanometers?

Ans. Ballistic galvanometer ~150 Ω

Dead beat galvanometer ~25 Ω.

20. How does an ammeter read the main current in a circuit without affecting it?

Ans. Actually, the ammeter is caliberated by direct comparision with a standard instrument, e.g. with an accurately caliberated potentiometer. The markings on the dial scale of an ammeter are made according to the values obtained by potentiometer.

21. What is figure of merit of a galvanometer?

Ans. It is that value of the current which produces a deflection of one division on the scale.

22. What is current sensitivity of a galvanometer?

Ans. The deflection per unit current is called as current sensitivity, i.e.

$$S_i = \theta / i = \frac{NAB}{C}$$

23. How can you increase the current sensitivity of the galvanometer?

Ans. Current sensitivity S_i can be increased by (*i*) in creasing the number of turns (N) in the coil (*ii*) increasing the effective area (A) of the coil, (*iii*) increasing the magnetic induction (B) and (*iv*) decreasing the restoring couple per unit twist (C).

EXPERIMENT 17

Object

To convert a galvanometer into voltmeter of a given range and to caliberate it.

Apparatus

Galvanometer, an accumulator, a voltmeter of some range (as given for conversion), high resistance box, a low resistance box, rheostat.

Theory

A voltmeter is used to measure potential difference between two points in an electrical circuit. A galvanometer can be converted into voltmeter of range V volts by placing resistance R in series with it. R is given by

$$R = \frac{V}{I_g} - G \qquad (1)$$

where G is galvanometer resistance and I_g is the current for full scale deflection of galvanometer.

Method

1. The galvanometer resistance G and I_g the current full scale deflection of galvanometer is determined using half deflection method as explained in experiment 16.

2. Using the values of I_g and G determine the resistance R for given range V volts using equation (1).
3. This resistance R when connected in series with the galvanometer makes it a voltmeter of range V volts.

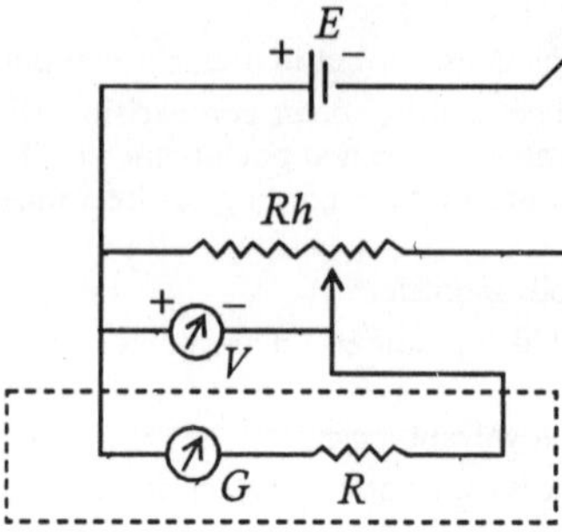

Fig. 34

4. For calibration make the connections as shown in Fig. 34.
5. Measure same potential difference with a voltmeter and with converted galvanometer (box in the diagram) and caliberate the whole range of the instrument.
6. Find error in potential difference as measured by converted instrument and the voltmeter.
7. Draw a caliberation curve, between voltmeter reading and corresponding converted galvanometer reading (Fig. 34).

Observations

Range of the voltmeter ... Volts.

(A) DETERMINATION OF G-GALVANOMETER RESISTANCE AND I_g

S.No.	*Resistance R (Ohms)*	*Galvanometer deflection n (Div.)*	*Resistance r for the deflection (Ohms)*	*Galvanometer resistance G = r (Ohms)*	$I_g = \frac{E}{R+G} \cdot \frac{N}{n}$ *(amp.)*

Mean I_g = ... amp.

(B) CALIBERATION OF CONVERTED GALVANOMETER

After conversion one division of galvanometer scale = ... volts.

S.No.	*Voltmeter reading (Volt) V*	*Galvanometer reading V'*		*Error (V − V') (Volts)*
		Div.	*Volts*	

Calculations

(i) $I_g = \frac{E}{R+G} \cdot \frac{N}{n}$ amps. (ii) $R = \frac{V}{I_g} - G$ Ohms.

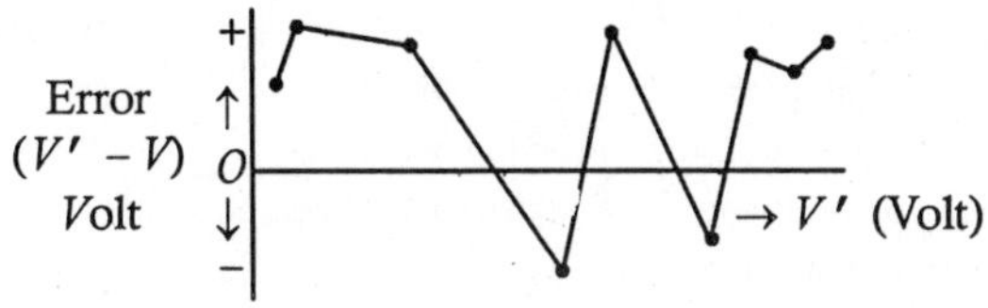

Fig. 35: Caliberation curve for a voltmeter

Results

1. The resistance R to be used in series with the galvanometer to convert it to voltmeter is ... Ohms.
2. Caliberation curve is as shown in graph (Fig. 35). The curve may be different for different voltmeters.

Precautions

Same as in experiment 16.

Questions with Answers

1. What is an voltmeter?

Ans. This is an instrument used to measure the potential difference between the two points in a circuit, directly in volts, when connected across these points in the circuit.

2. What should be the resistance of an ideal voltmeter?

Ans. Infinite.

3. Is it possible to use infinite resistance in an experiment?

Ans. No, in practice it is not possible. On connecting an infinite, i.e. very high resistance in series with a galvanometer, it won't give any deflection.

4. To overcome the above difficulty, what will you do?

Ans. Depending upon the range of a voltmeter, one can connect a high resistance with the coil of the galvanometer.

5. What is the aim of making resistance of a voltmeter high?

Ans. So that one can read the actual potential difference between the two points in a circuit. We may note that on connecting the voltmeter, this potential difference should not be affected. For this purpose, the current drawn by the voltmeter should be extremely small. This is why the resistance of a voltmeter is kept high.

6. How can a galvanometer be converted into a voltmeter?

Ans. For a particular range of voltmeter to be designated, first we find the value of high resistance that is to be connected in series with the galvanometer. Thus a galvanometer with high resistance in series acts as voltmeter.

7. How, one can increase the range of a voltmeter?

Ans. By increasing the value of series resistance.

8. What will be the extra resistance required to increase the range of a voltmeter of resistance 300 Ω form 3 V to12 V?

Ans. $R = (n - 1) R_v = (4 - 1) \times 300 = 900\ \Omega$.

9. Which instrument will you use for measuring alternating potential difference?

Ans. Hot-wire voltmeter.

10. Is it possible to change an ammeter into voltmeter?

Ans. Yes, by connecting a suitable high resistance in series with ammeter.

11. Can you convert an voltmeter into an ammeter?

Ans. Yes, by putting a suitable shunt of low resistance.

12. What will you do for accurate caliberation of an voltmeter?

Ans. We will use a potentiometer.

13. Is potentiometer an ideal instrument to measure potential difference?

Ans. Yes.

14. What is a millivoltmeter?

Ans. A low range voltmeter.

15. Is it possible to change a voltmeter into millivoltmeter?

Ans. Yes, by decreasing resistance of voltmeter.

EXPERIMENT 18

Object

To caliberate an ammeter with the help of a potentiometer.

Apparatus

Potentiometer, two rheostats, two keys, one two-way key, standard 2 Ω resistance, standard cell, galvanometer, two storage batteries and one ammeter.

Method

(*i*) Arrange the apparatus as shown in Fig. 36. *ABD* is potentiometer circuit. *S* is the standard cell (*e.m.f.* = 1.0183 V), *r* is a standard resistance coil of resistance of 2 Ω and *K* is a two-way key.

(*ii*) Allow to pass a small current, *i* through the ammeter by adjusting *R* and plugging in the key K_1. The potential difference at the ends of

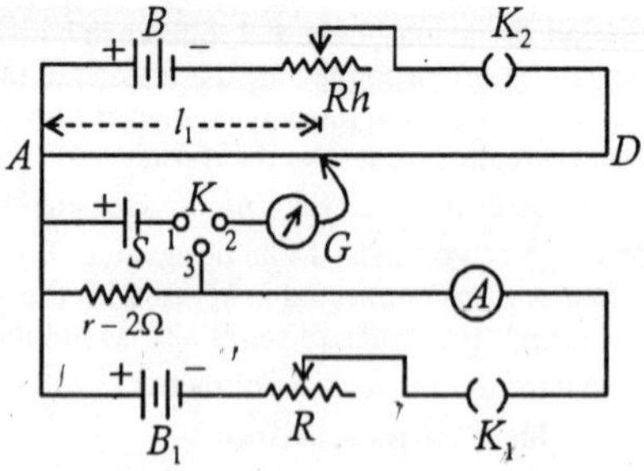

Fig. 36

r is $e = i.r$. Join 2 and 3 and obtain a balance point on potentiometer wire AD at a distance l_1 from A.

(*iii*) Open K_1 and join 1 and 2 in K and obtain a balance point for S at l_2 on AD. If e_s be the *e.m.f.* of S, $\frac{e}{e_s} = \frac{l_1}{l_2}$ or $e = ir = 1.0183 \times \frac{l_1}{l_2}$. or $i = \frac{1.0183}{r} \cdot \frac{l_1}{l_2}$. Compare this value with the ammeter reading. The difference is error.

(*iv*) Gradually reduce the resistance R, so as to increase i in suitable small steps and compare the ammeter-readings with the calculated values.

Observations

E. M. F. of the standard cell = 1.0183 volts.

S. No.	l_1(cm)	l_2(cm)	(*Calculated*) (amp)	*Ammeter reading* (amp)	*Difference* (amp)
1.	...	...	...	...	.
2.					
3.					
4.					

Calculations

$$i = \frac{1.0183}{r} \cdot \frac{l_1}{l_2}$$

Sources of Errors and Precautions

(*i*) Current should be passed through the various circuits for short intervals at a time, so as not to heat the wires and thus change their resistance.

(*ii*) All connections must be tight.

Experiment

To caliberate a voltmeter with the help of a potentiometer.

Apparatus

Potentiometer, two rheostats, two keys, one two-way key, standard two-ohm resistance, standard cell, galvanometer, storage batteries and a voltmeter.

Method

(*i*) Arrange the apparatus as shown in Fig. 37. *ABD* is the potentiometer circuit with *AD* as potentiometer wire. *S* is the standard cell (*e.m.f.* = 1.0183 V), r is a standard resistance of 2 Ω, and *K* is s two-way key.

(*ii*) Allow to pass a small current *i* through the circuit II, by keeping the rheostat resistance *R* large and plugging in the key K_1. Then potential difference at the ends of *r* is $e = ir$. Join 2 and 3 *K* and obtain a balance point at a distance l_1 from *A*.

(*iii*) Open K_1 and join 1 and 2 in *K* and obtain a balance point for *S* at l_2 on *AD*. If e_S be the *e.m.f.* of *S*, $\frac{e}{e_s} = \frac{l_1}{l_2}$, or $e = 1.0183\frac{l_1}{l_2}$.

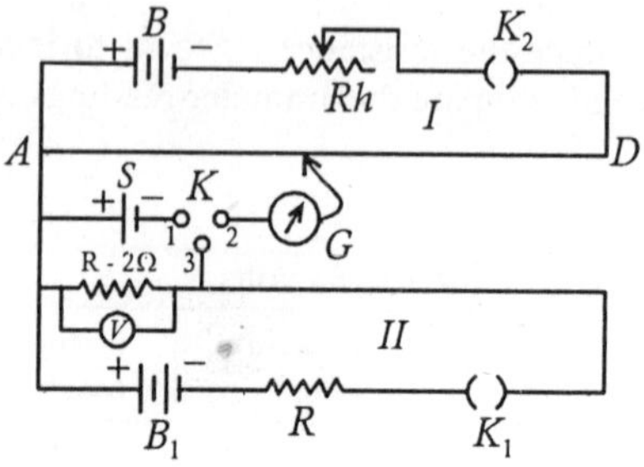

Fig. 37

Compare this values with the reading of the voltmeter *V*.

(*iv*) Gradually cut down the resistance *R* so as to increase the current *i* and, therefore, also the potential difference, *e* in suitable small steps.

Observation

As in previous experiment.

Calculations

$$e = 1.0183 \times \frac{l_1}{l_2}.$$

Sources of Error and Precautions

(*i*) The current should be passed through the various circuits for short interevals of time.

(*ii*) All connections must be tight.

EXPERIMENT 19

Object

To determine band gap in semiconductors using a P–N junction diode.

Apparatus

Germanium diode (0 *A* 5), microammeter (0 to 15 μA), a battery of 6 volt; a mercury thermometer with least count 0.1° C, a calorimeter with stirrer and carbon tetra chloride liquid.

Theory

In a semiconductor an energy gap exists between the top of the valence band and the bottom of the conduction band. Its electrical conductivity is governed by the number of electrons in the conduction band and holes in valence band. In a P—N junction diode the two carriers depend almost on the inpurity concentration but the conentration of minority carries, i.e., holes in N-material and electrons in P-meterial is dependent on the gap between the valence and conduction bands. When a junction diode is appreciabley reverse biased the reverse saturation current is due to these minority carriers only and its measurement as a function of temperature can be used to determine the band gap.

The characteristic of a junction diode is given by:

$$I = I_S[\exp(eV/kT) - 1] \tag{1}$$

where I is the current, V is the external voltage applied across the diode, k is Boltzmann constant and T is the absolute temperature. I_S the reverse saturation current which is given by

$$I_S = Ae(N_p V_n + P_n V_p) \tag{2}$$

where N_p and P_n are the concentrations of the electrons in the P-material and the holes in the N-material respectively. V_n and V_p are the mean velocities of the electrons and holes and A is the area of the junction.

If E_{CP}, E_{VP}, and E_{FP} represent the bottom of the conduction band, top of valency band and Fermi level in P-material and the corresponding levels in the N-material are represented by $E_{Cn} E_{Vn}$ and E_{Fn} as shown in Fig. 37. It can be shown that:

$$n_P = N_n \exp[-(E_{CP} - E_{FP})/kT] \tag{3}$$

$$p_p = N_p \exp[-(E_{VP} + E_{FP})/kT] \tag{4}$$

$$n_n = N_n \exp[-(E_{Cn} - E_{Fn})/kT] \tag{5}$$

$$p_n = N_p \exp[-(E_{Fn} - E_{Fn})/kT] \tag{6}$$

Here $$N_n = \frac{2(2\pi m_n kT/e)^{3/2}}{h^3}$$

and $$N_p = \frac{2(2\pi m_p kT/e)^{3/2}}{h^3}$$

m_n is the mass of electron and m_p effective mass of the hole and h is Planck's constant. As a result of the junction, the Fermi level will adjust so that

$$E_{FP} = E_{Fn}$$

or $$n_P p_P = n_n p_n = N_n N_p \exp[-\Delta E/kT] \tag{7}$$

where $$\Delta E = E_{CP} - E_{VP} = E_{Cn} - E_{Vn.}$$

At the temperature where the donor and acceptor leaves are fully activated, the density of electrons n_n in the n-meterial and the density of holes p_p in the p-material will be approximately equal to the impurity concentration and will be constant for small range of temperatures.

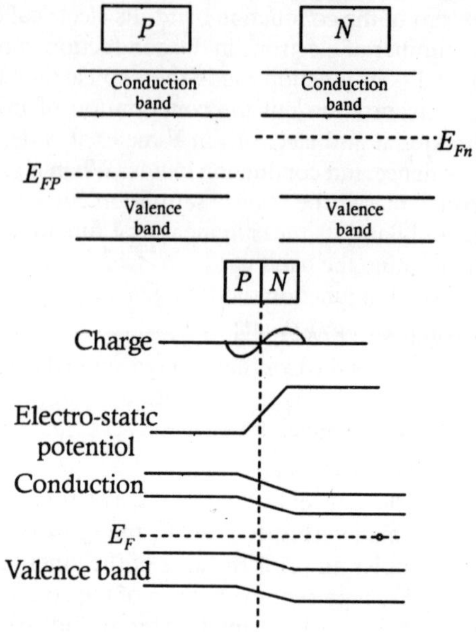

Fig. 38

Thus $$n_p = \frac{N_n N_p}{p_p} \exp\left[-\Delta E / kT\right] \tag{8}$$

and $$p_n = \frac{N_n N_p}{n_n} \exp\left[-\Delta E / kT\right] \tag{9}$$

Substituting the values of n_P and p_n in Eq. (2), we obtain

$$I_S = AeN_n N_P \left[\frac{V_n}{p_p} + \frac{V_P}{p_n}\right] \exp\left[-\Delta E / kT\right] \tag{10}$$

where the junction diode is appreciably reverse biased, the exponential term in Eq. (1) becomes small and $I = -I_S$.

Now, from Eq. (10), we obtain

$$\log_e I_s = \log\left[AeN_n N_P \left(\frac{V_n}{P_P} + \frac{V_P}{p_n}\right)\right] - \frac{\Delta E}{kT} \tag{11}$$

over a small range of temperature, the temperature dependence, of I_S in Eq. (10) is predominantly governed by the term containing the reciprocal of temperature. Hence a plot of I_S with I/T gives a straight line. The slope of this line can be used to determine the band gap ΔE.

Expressing logs to the base 10 and ΔE in eV (Eq. 11) can be written in the form

$$\log_{10} I_S = \text{const} - \frac{1.6 \times 10^{-19} \Delta E}{1.38 \times 10^{-23} \times 2.303} \cdot \frac{1}{T}$$

$$= \text{const} - 5.056 \times \left(\frac{10^3}{T}\right) \qquad (12)$$

Method

Make the connections as shown in Fig. 39 connecting a germanium diode in series with a micro-ammeter and battery.

Put this germanium diode in the calorimeter containing a stirrer, filled with liquid carbon tetra chloride heated upto 70° C and surrounded by a non-conducting material (cotton, wool, etc.) to keep the rate of cooling low.

Insert the thermometer in the calorimeter so that the bulb of thermometer remains close to diode.

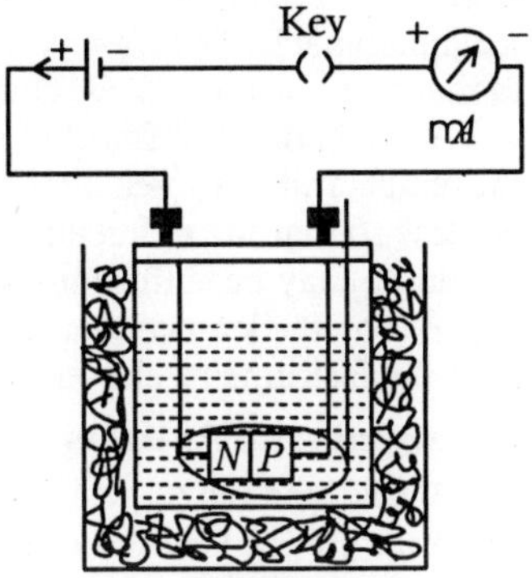

Fig. 39

Stir the liquid continuously and measure the current at various temperatures as the diode cools from 70° C to about room temperature.

Plot a graph between $\log_{10} I_S$ and (10^3/T).Calculate the band gap for geranium my knowing the slope of the graph obtained.

Observation

S.No.	Temperature $T°$ K	Current I_g amp.	$\frac{10^3}{T}$	$\log_{10} I_S$

Calculations

Plot a graph between (10^3/T) and $\log_{10} I_S$ which will be a straight line (Fig. 40) Find the slope ($\tan\alpha$) of the line from the graph and equate it to 5.303 ΔE.

$$5.303\, \Delta E = \tan\alpha$$

or $$\Delta E = \frac{\tan\alpha}{5.303} \text{ eV}$$

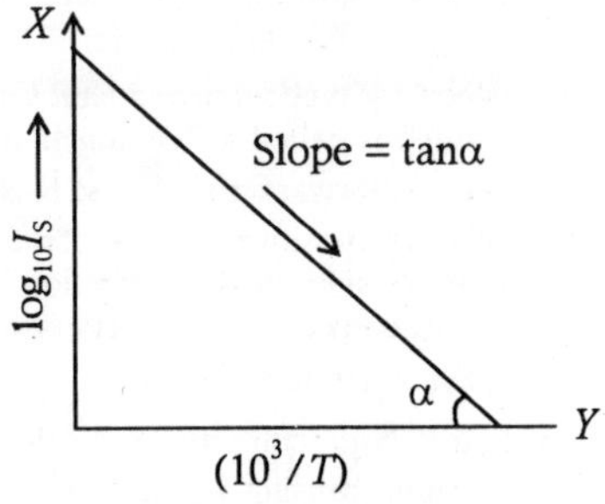

Fig. 40

Result

The band gap for given semiconductor is ... eV.
Its standard value is ... eV.

Precautions and Criticism

Carbon tetra chloride should be heated by keeping the copper calorimeter containing it in hot water. It should not be heated directly on the heater as it is inflammable. The crystal diode should be connected inverse bias because in that case *I* is taken to be equal to *I*.

P-N junction is enclosed in a cover as such it takes sometime to attain the temperature as given by the thermometer. This requires the cooling to be slow so that the difference in temperature obtained outside and inside the diode may be sufficiently small. The accuracy of the result obtained depends on the least counts of the thermometer and the microammeter as such the instruments chosen should satisfy this condition.

Questions with Answers

1. Define band gap.

Ans. Band gap is a forbidden energy gap between the valence band and conduction band of a crystal.

2. What are intrinsic and extrinsic semiconductors?

Ans. A pure semiconductor is called intrinsic semiconductor, e.g. germanium or silicon. When these are doped with impurities semiconductors are called extrinsic.

3. Why does a semiconductor behave like an insulator at 0 K?

Ans. At 0 K almost all electrons in the valence bond do not have enough energy to crossover the forbidden energy gap so as to reach the conduction band and to make themselves available for conduction of electricity. This is why semiconductors at 0 K behaves as an insulator.

4. What is a reverse current across a P-N junction?

Ans. When a P-N junction is reversed biased the current due to minority carries whose concentrations is dependent on energy gap or band gap is called reverse current.

5. What are P and N types semiconductors?

Ans. When a trivalent impurity like Al, Ga, B, In, etc. is added in 4th group elements, semiconductor is called P-type. When a pentavalent impurity P, As, Sb, B i, etc. is added to 4th group elements, the semiconductor is called N-type.

6. What is a P-N junction?

Ans. When a P-type semiconductor is used with a N-type semiconductor a junction formed is called a P–N junction diode.

7. What is forward and reverse biasing of a junction?

Ans. When the positive terminal of the battery is connected to the P-side and negative to the N-side, the P–N junction is said to be forward biasing. On reversing the connections either of battery or of the p-n junction, is said to be reverse biased.

8. What is a depletion layer?

Ans. In a P–N junction diode, the drift of the electrons and holes takes place at the junction, causing a potential barrier which stops further drift of electrons and holes. This layer is called depletion layer.

9. How do the impurities affect the conductivity of semiconductors?

Ans. Conductivity of the semiconductors increases.

10. What do you understand by majority and minority carriers?

Ans. In a P-type semiconductor holes are majority carriers and electrons are minority carriers. Whereas in N-type semiconductors electrons are majority carriers and holes are minority carriers.

11. What do you understand by forward and reverse resistance?

Ans. When a P-N junction diode is forward biased, then its resistance is called the forward resistance, which is of the order of 100 Ω. When the P-N diode is given reverse biase, then its resistance is called the reverse resistance, which is of the order of megaohms.

12. In the reverse biase arrangement on what factors do the reverse current depend?

Ans. (*i*) temperature and (*ii*) energy gap.

13. What are the orders of currents in forward and reverse biase arrangement of a P–N diode junction?

Ans. (*i*) forward bias current ~ milliampere

(*ii*) reverse bias current ~ microampere

EXPERIMENT 20

Object

To plot graph showing the variation of magnetic field with distance along the axis of a circular coil carrying current and to estimate from it the radius of the coil.

Apparatus

Stewart and Gee type tangent galvanometer, one storage battery, one rheostat, an ammeter, one commutator and one plug type key.

Description of the Instrument

The apparatus known as the Stewart and Gee type tangent galvanometer consists of a circular coil having a number of turns of insulated copper wire fixed with its plane vertical on a suitable horizontal bench B, and a magnetometer compass box which can slide on the bench such that the centre of the needle always lies on the axis of the coil (Fig. 41).

The distance of the needle from the centre of the coil can be read off with the help of linear scales SS provided on the two arms of the magnetometer.

Theory

The magnetic induction along the axis of current carrying circular coil is given by

$$B = \frac{\mu_0 n i a^2}{2\left(a^2 + x^2\right)^{3/2}} \qquad (1)$$

where a = radius of the coil,

n = number of turns,

i = current through the coil,

x = distance of the point along the axis from the centre of the coil.

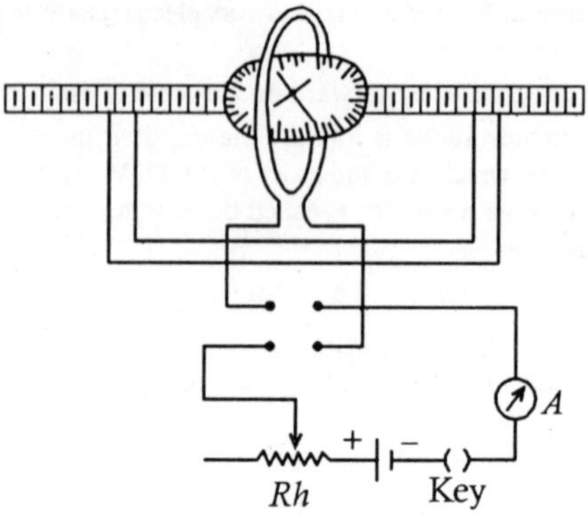

Fig. 41

If the instrument is adjusted such that this field is perpendicular to the horizontal component of the earth's magnetic induction (B_0), then

$$B = B_0 \tan\theta \quad (2)$$

where θ is the deflection of the small magnet placed at the point of observation. Hence the field due to coil can be taken proportional to the tangent of the deflection ($\tan\theta$) and gives the measure of the magnetic field.

If a graph is plotted with X as abcissa and $\tan\theta$ as ordinate, a curve symmetric about the Y-axis shall be obtained (Fig. 42). At the point where the curve changes sign, i.e. at the point of inflexion $a = 2x$. Hence distance between the two points of inflexion lying on the two branches of the curve gives the radius of the coil.

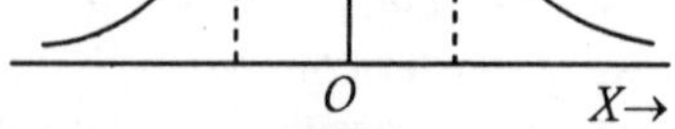

Fig. 42

Method

Place the instrument on the table in such a way that the arms of the sliding magnetometer point roughly east-west and the magnetic needle of the compass box is at the centre of the coil. By placing the eye a little above the coil, rotate instrument in the horizontal plane in such a way that the coil, the needle and its image all lie in the same vertical plane. This adjustment puts the coil roughly in magnetic meridian.

Now to adjust the coil exactly in the magnetic meridian make the electrical connections as shown in Fig. 41. With the help of rheostat adjust the current in the coil so that of deflection in the needle is of the order of 60°-65°. After lightly tapping the glass cover of the compass box with a finger, note the deflection, at the two ends of the points. Now reverse the current in the coil with the help of commutator and again note the deflection as before. If the mean deflection in the two cases does not agree closely, the coil is not in the magnetic meridian. Slightly turn the coil and repeat the process untill the mean deflections with the direct and reverse currents agree as closely as possible.

Now, for $x = 0$, i.e., when the needle is situated at the centre of the coil note the deflection both for direct and reverse currents and calculate the mean deflection θ. Now shift the compass box by 2 cm and note deflection as before. Be sure that the current till the current through the coil does not change.

Continue this process till the compass box reaches the end of the bench.

Repeat the experiment exactly in the same manner on the other side of the coil.

Observations

Number of turns in the coil, $n = \ldots$

Current through the coil, $i = \ldots$ amp

Radius of the coil $a = \ldots$ cm.

S. No.	Distance from the centre x	*Direct Current* One end of the pointer	Second end of the pointer	*Reversed Current* One end of the pointer	Second end of the pointer	Mean deflection θ	tan θ

Calculations

Plot a graph taking x-along X-axis and tan θ along the Y-axis. The curve will be symmetrical and will have maximum values when the needle is at the centre of the coil itself. Find the two points of *inflexion*, i.e. where the curve changes the curvature. Determine the distance between the two points which gives the radius.

Result

The graph showing the variation of the magnetic field along the axis of the given current carrying circular coil is attached here with (shape is shown in Fig. 42)

The radius of coil is equal to ... cm.

The radius by actual measurement is ... cm.

Precautions and Criticism

(*i*) The experiment should be conducted at a place where in its vicinity, there is no magnetic materials or current carrying conductors. For the same reason the rheostate of the circuit should be kept sufficiently far away from the compass box.

(*ii*) The Plane of the coil should be carefully adjusted in the magnetic meridian so that tangent law can be applied.

(*iii*) The current of the coil should be carefully adjusted in the magnetic meridian so that tangent law can be applied.

(*iv*) The current in the coil should be so adjusted that the deflection in the needle at the centre of the coil is nealy 75°. This current should remain constant throughout the experiment. For this purpose a storage battery of large capacity should be used. More over an ammeter should be included in the circuit and the current should be adjusted to a constant value with the help of rheostat.

(*v*) Both ends of the pointer should be read to avoid error due to the excentricity (i.e., the pivot of needle does not coincide with centre of circular scale). In order to avoid error due to parallax in reading the deflection, a plane mirror attached to the base of the compass box should be used.

(*vi*) In order to avoid friction effects at the pivot the readings should be taken after gently tapping the glass top of the compass box with finger.

(*vii*) Deflection of the needle should be recorded both for direct and reverse currents.

The main source of errors are: the coil may not be exactly circular and needle may not be situated exactly at its centre.

The plane of the coil may not be in magnetic meridian.

The magnaetic needle is not very small and its own field causes the error and tangent law is not accurately obeyed. The friction at the pivot may not be totally absent. The scale and pointer method for measuring deflection is not very accurate.

Questions with Answers

1. What is the direction of the magnetic field at the centre of the coil?

Ans. Normal to the plane.

2. Where do you expect the magnetic field to be maximum in this experiment?

Ans. At the centre of the coil.

3. Explain the behaviour of the field near the centre of the coil.

Ans. Near the centre, the field is uniform, i.e. the lines of force due to the current in a small region round about the centre are equidistant and parallel.

4. What is the use of an extra pointer in the compass box?

Ans. A long light pointer is used to increase the sensitivity in measuring θ .

5. What is magnetic meridian?

Ans. Magnetic meridian is an imaginary vertical plane passing through the magnetic axis (N-S pole joining line) of a freely suspended magnet.

6. How will you ascertain that the coil is in magnetic meridian?

Ans. By reversing the current in the coil, if the deflection due to direct and reverse currents are identical, the coil is said to be set in magnetic meridian.

7. Suppose the coil is not set in the magnetic meridian, what error will be caused by it?

Ans. One can not apply tangent law.

8. What is point of inflexion?

Ans. The point where the curve changes its curvature is called point of inflexion.

9. Can you obtain uniform magnetic field within a large space?

Ans. Yes, by placing two identical coils parallel to each other with their centres at a distance equal to their radius.

10. Why a plane mirror placed at the base of the compass box?

Ans. To avoid parallax, while measuring deflection θ .

11. What do you understand by magnetic induction?

Ans. The number of magnetic lines of force passing through unit normal area at any point in a magnetic field is called magnetic induction (B) at that point.

12. What is the unit of magnetic induction in S.I. units?

Ans. Wb/m^2 or Tesla or Newton/Amp-m.

13. How does a small coil carrying current behave?

Ans. Like a small dipole. The magnetic moment is equal to the product of the current flowing in the coil and the effective area of the coil.

EXPERIMENT 21

Object

To determine high resistance by method of leakage of a condenser with the help of a condenser and ballistic galvanometer.

Apparatus

A condenser (~1 μF), high resistance (~mega ohms), accumulator, ballistic galvanometer with lamp and scale arrangement, stop watch, morse key, a two way key and a one way key.

Theory

If a fully charged condenser (initial charge θ_0) of capacity C is connected to a high resistance R, the charge leaks through the resistance and at any instant t the residual charge on the condenser is given by

$$Q = Q_0 e^{-t/RC} \qquad (1)$$

Equation (1) shows that the discharge of a condenser is not instantaneous but takes some time which depends on the time constant RC of the circuit. If R is very high, discharge takes place very slowly and hence the time of discharge can be measured accurately.

Let a condenser of capacity C is given a charge Q_0 and discharge immediately through a ballistic galvanometer. If θ_1 is the first throw of the galvanometer and λ the logarithmic decrement, then

$$Q = K\theta_1 (1 + \lambda/2) \tag{2}$$

Now let the same condenser be again given the same charge and after having allowed the charge to leak through resistance R for a known time t is discharged through the same galvanometer. If ϕ_1 is the first throw of the galvanometer, then

$$Q = K\phi_1 (1 + \lambda/2) \tag{3}$$

When Q is the charge left on the condenser after time t.

$$\therefore \quad \frac{Q}{Q_0} = \frac{\phi_1}{\theta_1} = e^{-t/RC}$$

$$\therefore \quad R = \frac{t}{\log_e(\theta_1 / \phi_1)} = \frac{t}{2.303 \log_{10}(\theta_1 / \phi_1)} \tag{4}$$

This equation is used to determine high resistance. Re-arranging the terms, we have

$$\log_0(\theta_1 / \phi_1) = \frac{t}{2.303\,RC}$$

i.e., a graph between $\log_{10} (\theta_1/\phi)$ and t is a straight line, the slope of which is given by

$$\tan\alpha = \frac{1}{2.303\,RC} \quad \text{or} \quad R = \frac{1}{2.303\ C\ \tan\alpha} \tag{5}$$

The above calculation is true only when the charge on the condenser does not leak through its dielectric. If the condenser has a self-leakage resistance (r), then a connection has to be made. For this purpose, the experiment is performed in two parts.

(*i*) The charge is allowed to leak with the external resistance R connected across the condenser. Then Eq. (4) and (5) will give (R) the equivalent resistance of r and R' connected in parallel, then

$$\frac{1}{R} = \frac{1}{R'} + \frac{1}{r} \tag{6}$$

Method

Adjust the galvanometer properly so that the coil moves freely between the pole pieces of the magnet and adjust the lamp and scale arrangement to get a bright spot of light reflected from the galvanometer mirror on the scale. Bring the spot of light on the zero of the scale.

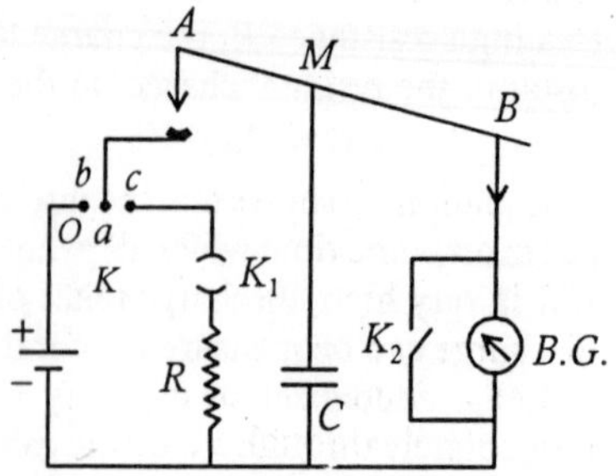

Fig. 43

Now make the electrical connections as shown in Fig. 43. M is

Morse key, K is two way key, whereas K_1 and K_2 are the plug-key and the tapping key-respectively, C is the condenser of known capacity and R is unknown high resistance.

Connect a to b and then press the end A of the Morse key. In this process the condenser plates are connected with the cell. When the condenser gets fully charged release the end A. In this process the cell circuit is broken with the condenser, which is now connected with the ballistic galvanometer. The condenser gets discharged through the galvanometer. Note down the first throw of the galvanometer θ_1.

Charge the condenser again to the same voltage by depressing the end A of the Morse key. Now keeping the end A depressed and also the plug key K_1 open, turn the key K from '*ab*' to '*ac*' position and immediately start a stop watch. In this case the condenser will discharge through its dielectric. Let it discharge for a known time t. Now release the Morse key so that condenser is discharged through the galvanometer and note the throw (ϕ'_1) of the galvanometer. Calculate the value of the self-leakage resistance r using Eq. (4).

Now close the plug key K_1 and turn the key K from '*ac*' to '*ab*' position and depress the end A. When the condenser is fully charged turn the key k from '*ab*' position to the '*ac*' position at the same time keeping the end A pressed start the stop watch. Now the condenser will discharge through the high resistance R as well as through its own dielectric. Let the discharge for a known time t. Release the Morse key and note the first throw (ϕ_1) of the galvanometer. Calculate R' and R.

Repeat the experiment for different discharge timings of the condenser.

Observations

Capacity of the condenser C = ... μF

Rest position of the spot = ... zero

First throw of the galvanometer corresponding to full charge on it = ... cm.

S. No.	*Leakage time t*	*First throw of the gal. after charge has leaked through*		$\log_{10}$ (θ_1/ϕ'_1)	$\log_{10}$ (θ_1/ϕ_1)
		self leakage resistance r of the condenser (ϕ_1)	*parallel combination of r and R* (ϕ_1)		

Calculations

Plot a graph between t and $\log_{10}(\theta_1/\phi'_1)$ and between t and $\log_{10}(\theta_1/\phi'_1)$. Both are straight lines. From the graph between t and $\log_{10}\left(\frac{\theta_1}{\phi'_1}\right)$, we have

$$r = \frac{1}{2.303\,C\tan\alpha'} = \ldots\ \Omega$$

From the graph between t and $\log_{10}(\theta_1/\phi_1)$ we have

$$R' = \frac{1}{2.303\,C\tan\alpha} = \ldots\ \Omega$$

Value of high resistance $R = \frac{r - R'}{R'r} = \ldots\ \Omega$

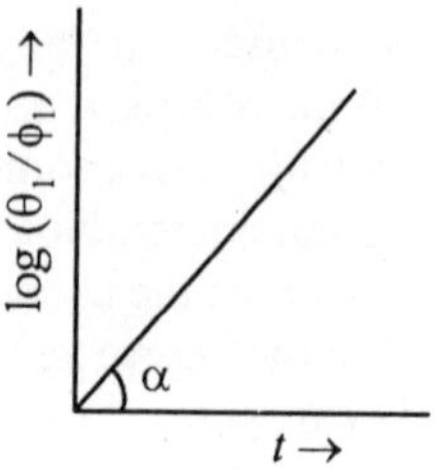

Fig. 44

Result

The value of the given high resistance determined by the method of leakage of condenser = ... Ω.

Precautions

1. The experiment if performed carefully gives accurate results. The method is however suitable for determining high resistance of the order of several megahms. The method can be used to measure the resistance and the capacity of condenser if a suitable high resistance is available.
2. There should be no leakage of charges through the keys.
3. Accumulator should be used for charging the condenser.
4. The capacity of the condenser should be such as to ensure a satisfactory rate of leak.
5. All the connecting wires should be kept away from each other. They should not touch the experimental table or the instrument case.

Questions with Answers

1. What is a condenser?

Ans. Condenser can store a large amount of charge. Usually it consists of two parallel metal plates placed at a short distance. The intermediate space between the plates is filled with some dielectric.

2. What is the use of a dielectric?

Ans. It increases the capacity of a condenser.

3. On what factors does the capacity of a condenser depend?

Ans. Capacity of a condenser is directly proportional to the area of the plates and the dielectric constant of the medium, and inversely proportional to the plate separation.

4. What is the necessity of charging a condenser each time for about half an hour?

Ans. A condenser when put to charging take some time to reach maximum value of charge (say q_0). Thereafter, this tends to oppose the introduction of further charging to the condenser from the cell or battery, i.e. a back e.m.f. sets up in the circuit. To overcome the back e.m.f. the main current takes some time.

5. Does a condenser get totally discharged immediately after connecting it with resistance in a circuit?

Ans. No, the discharge of a condenser is not instantaneous and it takes place in accordance with the relation $Q = Q_0 \exp[-t/RC]$, i.e. the discharge of a condenser is not instantaneous and it will take infinite time to get fully discharged.

6. What is time constant for the discharge of a condenser?

Ans. This is defined as that time taken by condenser to reduce its charge to 37% of the initial maximum value.

7. Can you measure high resistance of the order of megaohms, i.e. $10^6\ \Omega$ with this method?

Ans. Yes.

8. Why a condenser does not achieve its maximum charge capacity at once when the circuit is closed?

Ans. We know that charge on the condenser increasing exponentially and attains maximum value at infinite time. At any time t, the charge Q on the condenser is given by $Q = Q_0 (1 - e^{-t/Rc})$

where Q_0 is the maximum charge.

9. During experiment why observations are taken in two parts?

Ans. In order to take into account for the self leakage of the condenser through its dielectric.

10. What factors in a circuit affects the growth of the charge on the condenser?

Ans. (*i*) Resistance R and (*ii*) condenser C.

11. What is time constant of a circuit?

Ans. Time constant = RC.

12. Is in this experiment, should the possible leakage through the dielectric of the condenser be taken in account?

Ans. Yes, because the dielectric used may not be perfectly insulated and it is possible that some charge may leak out through the condenser even without connecting to resistance R. This is termed as self leakage resistance of the condenser.

13. Can you measure the high resistance of the order of $10^6\Omega$ by meter bridge or P.O. box?

Ans. No, these methods based on Wheatstone bridge principle are only suitable when the resistances of the four arms of the bridge are of the same order.

EXPERIMENT 22

Object

To determine the ballistic constant K of a moving coil ballistic galvanometer with a standard condenser of known capacity.

Ballistic Galvanometer

Ballistic galvanometer is used to measure charge, viz. discharge of a condenser that passes through it in a short duration. The special feature of this galvanometer is that the moment of inertia of the moving system is large so that the coil takes longer time to vibrate in the magnetic field. Time period T is given by

$$T = 2\pi\sqrt{\frac{I}{C}}$$

The time period T can be increased by decreasing the couple per unit twist C of the suspension fibre by using phosphor bronze wire. Further, in this galvanometer, the coil is wound on a non-conducting frame so that the damping due to eddy currents is reduced, and for the residual damping that exists, damping correction is employed.

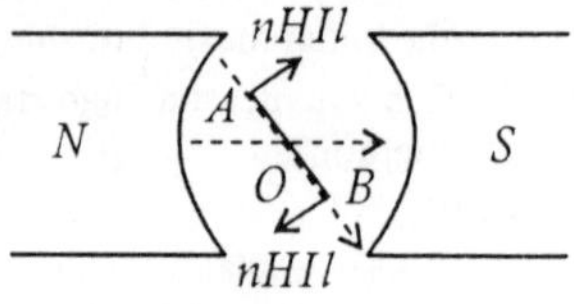

Fig. 45

If a current I is passed through the coil of n turns (length l, breadth b) and if the intensity of magnetic field is H, then force experienced by the length side of the conductor of coil will be $nHIldt$. If the current flows for a small interval of time dt_1 then,

Impulse of the force $= nHIldt$

$= \text{charge in momentum}$

Change in momentum for a charge q that flows through the coil

$$= \int nHI\,l\,dt = nHl \int Idt = nHlq$$

$\therefore$ Angular momentum $= nHql \times b = nHqA$

where A is the face area coil ($A = l \times b$).

But angular momentum $= I\omega$. Here I is the moment of inertia of the moving system about the axis of suspension and ω is angular velocity. We have

$$\therefore \quad nHqA = I\omega \qquad (1)$$

Due to angular velocity ω, the kinetic energy of the oscillating system is $\frac{1}{2}I\omega^2$. This energy is used up in twisting the suspension fibre through an angle θ. If C is the couple per unit twist, then work done in twisting the suspension fibre through on angle θ is $\frac{1}{2}C\theta^2$

$$\therefore \quad \frac{1}{2}I\omega^2 = \frac{1}{2}C\theta^2 \qquad (2)$$

But $$T = 2\pi\sqrt{\frac{1}{C}}$$

$$\therefore \quad I = \frac{CT^2}{4\pi^2} \qquad (3)$$

From (2) and (3)

$$I^2\omega^2 = \frac{C^2T^2\theta^2}{4\pi^2} \quad \text{or} \quad I\omega = \frac{CT\theta}{2\pi} \tag{4}$$

Equating Eqs. (1) and (4), we obtain

$$q = \frac{T}{2\pi}\cdot\frac{C}{nAH}.\theta \tag{5}$$

$$= K\theta \tag{6}$$

where K is a constant and known as ballistic constant, which may be defined as the quantity of charge required to produce a unit angular deflection of the coil. Here θ is the observed throw in the ballistic galvanometer.

In the above formulation (Eq. 6), it has been assumed that the damping experienced by the moving system is negligibly small. Hence for any damping which still exists, a correction has to made. If θ_1, θ_2, θ_3 ... be the successive swings to left and right (Fig. 46), it is found that

$$\frac{\theta_1}{\theta_2} = \frac{\theta_2}{\theta_3} = \ldots = \text{a constant} = d\,(\text{say})$$

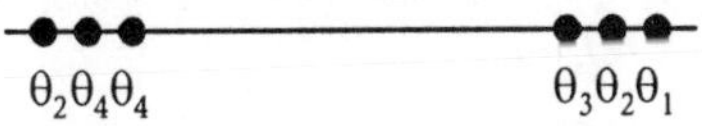

Fig. 46: Representation of successive swings

The constant d is called the decrement, and $\log_e d$ is called the logarithmic decrement and is denoted by λ. Thus

$$\log_e d = \lambda \quad \text{or} \quad d = e^\lambda$$

Hence
$$\frac{\theta_1}{\theta_2} = \frac{\theta_2}{\theta_3} = \ldots \quad \ldots = e^\lambda$$

The decrease from θ_1 to θ_2 takes place in half vibration and $\frac{\theta_1}{\theta_2} = e^\lambda$.

Clearly for a whole vibration $\frac{\theta_1}{\theta_3} = \frac{\theta_1}{\theta_2}\cdot\frac{\theta_2}{\theta_3} = e^{2\lambda}$ and so on. Thus if θ_1 be observed first swing and θ what it would have been if damping were absent, then, as the period in question is a quarter vibration.

$$\frac{\theta}{\theta_1} = e^{\lambda/2} = 1 + \lambda/2 + \text{terms in higher powers of } \lambda$$

$$= 1 + \lambda/2 \text{ (since } \lambda \text{ is small)}$$

$$\therefore \qquad \theta = \theta_1(1 + \lambda/2) \tag{7}$$

Thus, to correct for damping the observed first swing θ_1 of the moving system should be multiplied by the factor $(1 + \lambda/2)$.

To evaluate the logarithmic decrement, we see that

$$\frac{\theta_1}{\theta_2} = \frac{\theta_2}{\theta_3} = \ldots = \frac{\theta_{10}}{\theta_{11}} = e^{\lambda}$$

or $$\frac{\theta_1}{\theta_{11}} = e^{10\lambda}$$

$\therefore$ $$\lambda = \frac{1}{10}\log_e \frac{\theta_1}{\theta_{11}}$$

$$= 2.303 \times \frac{1}{10}\log_{10}\frac{\theta_1}{\theta_{11}} \quad (8)$$

Using Eq. (7), and (6) takes the form

$$q = K\theta_1(1 + \lambda/2) \quad (9)$$

where λ is given by eq. (8).

1. Measurement of K Using Standard Condenser

In order to determine the ballistic constant K of the galvanometer, a condenser of known capacitance (C) is first charged to a potential difference of E volts (by connecting it across an accumulator and then discharged through the ballistic galvanometer). Thus the charge on the condenser

$$q = CE = K\theta_1(1 + \lambda/2)$$

or $$K = \frac{EC}{\theta_1(1+\lambda/2)} \quad (10)$$

θ_1 is the throw of the galvanometer coil and Eq. (10) is the required rotation.

2. Measurement of K Using Constant Deflection Method

According to Eq. (6) given above

$$q = K\theta, \text{ where } K = \frac{C}{nAH}\cdot\frac{T}{2\pi}$$

If a steady current I produces a steady deflection ϕ in the coil, we have

$$n\,HIA = C\phi$$

or $$\frac{C}{nAH} = \frac{I}{\phi}$$

or $$\frac{C}{nAH}\cdot\frac{T}{2\pi} = \frac{T}{2\pi}\left(\frac{I}{\phi}\right) = K \text{ form eq. (6)} \quad (11)$$

For producing a steady current, a potential divider arrangement is employed as shown in Fig. 47. If E the e.m.f. on the cell is distributed over the total resistance $(R_1 + R_2)$, the refraction of the e.m.f. over the resistance R_1 is give by $\frac{ER_1}{R_1 + R_2}$.

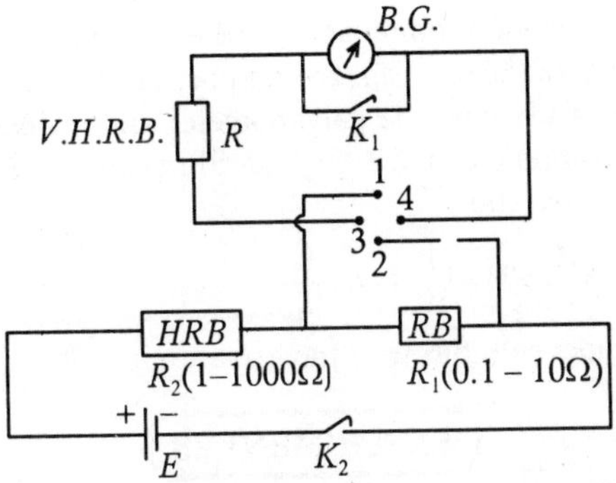

Fig. 47: Ballistic constant by constant deflection method

This e.m.f. is utilized to send the steady current I through the galvanometer, and a high resistance R (of the order of 10,000 Ω). The current I flowing through the galvanometer circuit is given by

$$I = \frac{R_1 E}{(R_1 + R_2)(R + G)}$$

For $R >> G$, we have

$$I = \frac{R_1 E}{R(R_1 + R_2)} \tag{12}$$

This value of I and the corresponding deflection produced by it are substituted in Eq. (11) to get the value of K. The time period is observed in the usual manner by means of a stop watch.

3. *Measurement of Ballistic Constant (K) by Means of an Earth Inductor*

Connect an earth inductor across the ballistic galvanometer (Fig. 48). The earth inductor is so arranged that its axis is vertical and its plane is perpendicular to the magnetic meridian. If the coil is rotated quickly through 180° and the first and successive throws of galvanometer coil are noted, then the ballistic constant K, of the galvanometer is given by the formula,

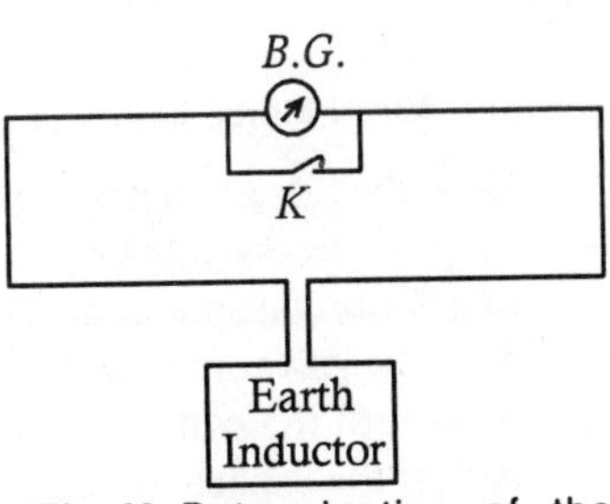

Fig. 48: Determination of the ballistic constant (K).

$$K = \frac{2NAH}{R\theta_1(1 + \lambda / 2)} \tag{13}$$

where H is the horizontal component of the earth's magnetic field, A is the effective face area of the coil, N is the number of turns in the coil, R is the total resistance of the coil, the galvanometer and the leads, θ_1 the first throw of the galvanometer and λ the logarithmic decrement which can be obtained from the relation

$$\lambda = 2.303 \times \frac{1}{10} \log_{10} \frac{\theta_1}{\theta_{11}}$$

where θ_{11} is the eleventh throw of the galvanometer.

EXPERIMENT 23

Object

To determine the ballistic constant of a moving coil ballistic galvanometer with the help of a condenser of known capacity.

Apparatus

Ballistic galvanometer, an accumulator, a capacity box, voltmeter, a morse key, a tapping key, etc.

Procedure

1. Make the connections as shown in Fig. 48

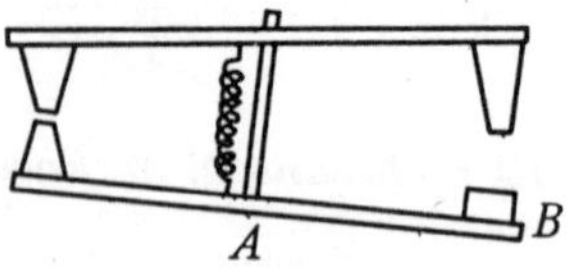

Fig. 49(*a*): Morse key

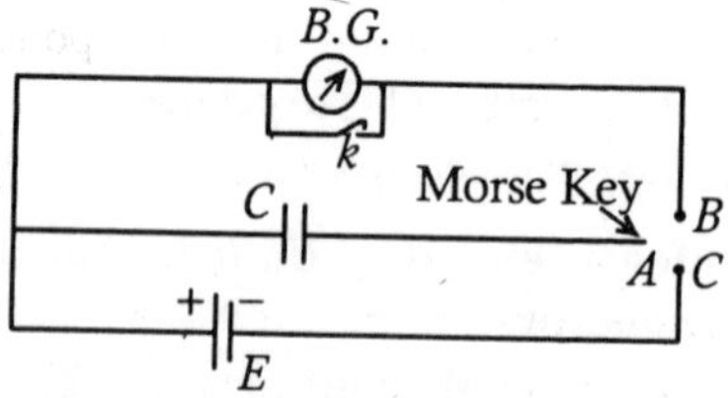

Fig. 49(*b*): Determination of Ballistic constant.

2. Level the galvanometer properly so that the coil moves freely between the pole pieces of the magnet. Adjust the lamp and scale arrangement to get a bright spot at the centre of the scale, after getting reflected from the galvanometer mirror. A damping key is used to check the motion of the coil.
3. Adjust the capacity box to a suitable value (say 0.1 to 0.5 μF).
4. Press the morse key to establish the contact between A and B. This charges the condenser to the value of the potential equal to the e.m.f. of the accumulator.
5. Now release the key (after 10 to 15 sec.), disconnects A from B and connect the condenser across the galvanometer. The condenser discharges sending its charge $q = CE$ through the ballistic

galvanometer. Note the first and the successive deflection of the galvanometer to find θ_1 and λ.

6. Repeat the process for different values of the capacity. Measure the e.m.f. (E) of the cell with a high resistance voltmeter. As the throw is sudden, it should be observed carefully. Calculate the mean value of K.

Observation

1. *E. M. F.* (E) of the accumulator used = ...volts.
2. Distance of the scale from galvanometer = ... cm.
3. Table for observed throws and λ :

S.No.	*Capacity of the condenser* μ F	*Deflection of the galvanometer coil*		$\frac{C}{\theta_1}$ *for each observation*	*Mean* $\frac{C}{\theta_1}$
		The first throw in mm. θ_1	*The eleventh throw in mm* θ_{11}		
1.					
2.					
3.					

Calculations

Set 1:

$$\lambda = 2.303 \times \frac{1}{10} \log_{10} \frac{\theta_1}{\theta_{11}}$$

$$= \ldots \ldots \ldots$$

Now $$K = \frac{EC}{\theta_1(1 + \lambda / 2)}.$$

$$= \ldots \ldots \ldots \text{ Coulomb/mm.}$$

Calculate K for other sets in the same way.-

Results

The ballistic constant of the given galvanometer = ... Coulomb/mm, when scale is kept at a distance of one metre from the galvanometer.

Precautions

1. The coil of galvanometer should be levelled properly so as to oscillate freely between the pole pieces of the magnet.
2. A tapping key should be connected across the galvanometer terminals so that undue swinging motion of the coil is immediately checked.
3. Always use an accumulator for giving constant e.m.f.
4. The capacitance of the condensers chosen should be of such a magnitude as to give suitable throw of light-spot on the scale. In

case spot moves off the scale, the potential divider arrangement should be used to charge the condenser.

5. The galvanometer should be clamped when not in use.
6. While using lamp and scale arrangement for measuring deflection, the scale should be placed one metre away from the galvanometer mirror.

Questions with Answers

1. What is constant of a ballistic galvanometer?

Ans. The constant of a ballistic galvanometer is defined as the charge per unit deflection (mm).

2. Name various types of damping in a ballistic galvanometer.

Ans. (*i*) Mechanical damping and (*ii*) electromagnetic damping.

3. What is difference between ballistic galvanometer and moving coil dead beat galvanometer?

Ans. In a dead beat galvanometer, the coil is wound on a conducting material, whereas in the case of ballistic galvanometer it is wound on a non-conducting frame, e.g. mica, paper, etc.

4. What is electromagnetic damping and how it is minimised?

Ans. The motion of the coil is opposed by inducedcurrent, when it moves in a strong magnetic field and this causes electromagnetic damping. One can minimise it by winding the coil on a non-conducting frame.

5. On what factors does the value of ballistic constant K depend?

Ans.
$$K = \frac{T}{2\pi}\frac{C}{NAB}$$

Obviously K is directly proportional to time period of the coil (T) and torsional constant (C) of the suspension fibre and K is inversely proportional to the magnetic field strength (B), number of turns (N) and area (A) of the coil.

6. Does the value of K depend on the working conditions and also on the distance between the scale and mirror?

Ans. When the measurement of deflection is made in terms of degrees, it does not affect K. However, when it is measured in mm, it affects its measurement. The working conditions also affect the result.

7. Can you determine K by any other method?

Ans. Yes, one can determine K by constant deflection method or by earth inductor.

8. What is logarithmic decrement?

Ans. Due to damping the throw of galvanometer is less than what it would have been in the absence of damping. After allowing the passage of charge, the deflection reduces in the following manner

$$\frac{\theta_1}{\theta_2} = \frac{\theta_2}{\theta_3} = \frac{\theta_3}{\theta_4} = \ldots\, e^{\lambda}$$

We call λ as the logarithmic decrement and it is given by

$$\lambda = \frac{2.3026}{10}\log_{10}\left(\frac{\theta_1}{\theta_{11}}\right)$$

9. Does K depend on the earth's magnetic field?

Ans. No.

10. What is current sensitivity and charge sensitivity?

Ans. Current sensitivity is defined as the current required to produce unit deflection in steady state. Charge sensitivity is defined as the charge required to produce unit deflection.

11. Mention some uses of ballistic galvanometer?

Ans. One can determine high resistance by leakage, magnetic field and capacity of a condenser using a ballistic galvanometer.

12. What is critical damping?

Ans. The damping at which transition from oscillatory to dead beat take place is known as critical damping.

13. Show that the relation between current sensitivity (S_i) and charge sensitivity (S_q)

Ans.
$$S_q = \frac{T}{2\pi} S_i$$

14. What is the relation between K and S_q?

Ans.
$$K = \frac{1}{S_q} \frac{\text{microcoulomb}}{\text{mm}}$$

15. Why the moment of inertia of a moving system in a ballistic galvanometer is made high?

Ans. Since the ballistic galvanometer is used to measure transient currents and therefore the moving part experiences a sudden impulse. If the duration of impulse is quite small compared to a quarter of its period, then whole of the charge passes through the galvanometer prior to its movable part has moved sensibly from its zero position, the magnitude of swing is independent of the time of duration of the impulse. Clearly, in order to increase the time period of the coil its moment of inertia is made high.

16. What is suspension fibre made of?

Ans. Usually, the suspension fibre is made of phosphor bronze, an alloy of copper (92.5%), tin (7%) and phosphorous (0.5%)

17. What type of mirror is used in the galvanometer?

Ans. Either concave or plane. If a plane mirror is used a convex lens has also to be used with it placed on glass cover on the galvanometer, in order to focus light on the mirror.

EXPERIMENT 24

Object

To determine the specific resistance of a given wire with the help of Carey Foster's bridge.

Apparatus

Carey Foster's bridge, galvanometer, Lechlanche cell, fractional resistance box, copper strip, plug key, rheostate (or 2 coils of equal resistances of either 1 or 2 ohms), 2 unknown resistance wires.

Theory

A Carey Foster's bridge is a modification of meter bridge in the following two ways.

(*i*) The effective length of the bridge wire is increased due to two additional gaps P and Q.

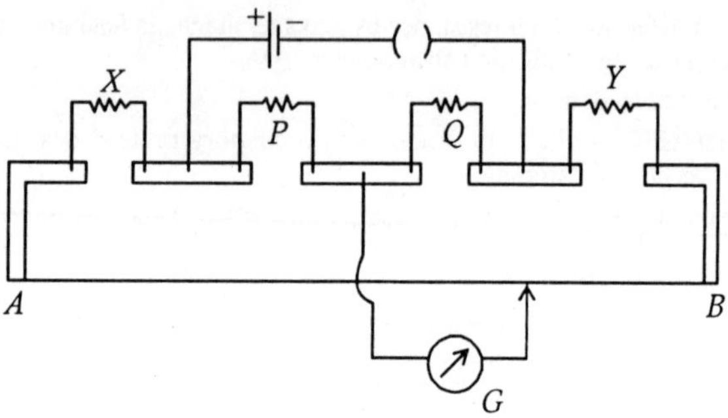

Fig. 50

(*ii*) The end resistances are eliminated by inter-changing the outer gaps. With these modifications the bridge is more precise and sensitive. Let ρ be the resistance per unit length of the bridge wire and α, β be the end resistance on two ends A and B respectively. If the balance is obtained at a length l_1 with X in L.H.S. gap and l_2 with X in R.H.S. gap, we have for the same value of ratio arms.

$$\frac{P}{Q} = \frac{X + \alpha + \rho l_2}{Y + \beta + \rho(100 - l_1)} \quad \text{for } X \text{ in L.H.S. gap}$$

and
$$\frac{P}{Q} = \frac{Y + \alpha + \rho l_1}{X + \beta + \rho(100 - l_2)} \quad \text{for } X \text{ in R.H.S. gap}$$

Solving, we obtain

$$X - Y = \rho(l_2 - l_1) \ldots \tag{1}$$

The equation gives the difference of two resistance in terms of bridge wire resistance.

To determine ρ, Y is kept zero, i.e., the gap is covered with copper strip and balancing lengths l_1 and l_2 are obtained for known values of resistance X. Then

$$\rho = \frac{X}{l_2 - l_1} \tag{2}$$

If X is known resistance, the unknown resistance Y can be determined.

$$Y = X - \rho(l_2 - l_1)$$

Knowing the resistance of wire, the specific resistance of its material can be determined as

$$\text{Specific resistance } \sigma = \frac{Ya}{l} = \frac{Y\pi r^2}{l} \qquad (3)$$

Where l is the length of the wire, r its radius.

Method

(A) Determination of ρ

1. Make neat and tight connections. A deci-ohm resistance box is connected in left gap and short circuit unknown gap on right by copper strip, Fig. 50.
2. Introduce some resistance and determine balance point by sliding Jockey. Choose such a resistance that balance occur towards the end of the wire.
3. Interchange resistance box and copper strip and find balancing length l_2 measured from the same end (i.e., left).
4. By changing the value of resistance X-in the resistance box, find the balancing length in such a way that whole length of the wire is caliberated.
5. Calculate the value of ρ using Eq. (2).

(B) Determination of Unknown Resistance and Difference between two Resistance

1. Use a known resistance box for X' having resistance comparable to unknown resistance Y connected in other gap.
2. By choosing suitable resistance from resistance box and balancing length l_1. By suitable we mean a resistance for which balance is obtained approximately at the centre of the bridge wire.
3. Interchange X and Y and find balancing length l_2.
4. By changing resistance X in steps and other sets of l_1 and l_2.
5. Calculate the difference (X-Y) from each of the above set.
6. Finally calculate Y the unknown resistance using Eq. (1).
7. For finding the different sets of observations for difference (X-Y) use rheostat for gaps P and Q.

(C) Determination of Specific Resistance

1. Find the length of resistance wire used.
2. Find the radius of wire using screw gauge.
3. Find specific resistance using Eq. (3)

Observations

(A) Determination of ρ, the Resistance Per Unit Length of Bridge Wire

S. No.	Resistance X (ohms)	Balancing length When resistance X is in		$(l_2 - l_1)$ (cms)	$\rho = \frac{X}{l_2 - l_1}$
		Left gap (l_1) cm	Right gap (l_2) cm		
1.					
2.					
3.					

(B) Determination of (X-Y), the Difference of Resistance and Determination of Unknown Resistance (Y) if X is Known

S. No.	Resistance X (ohms)	Balancing length When resistance X is in		$(l_2 - l_1)$ (cm)	$X - Y = \rho(l_2 - l_1)$	Unknown resistance $Y = X - \rho (l_2 - l_1)$
		Left gap l_1 (cm)	Right gap l_2 (cm)			

Mean $Y = \ldots$

(C) Determination of Radius of the Wire

Least count of the screw gauge=

Zero correction =

S. No.	M.S.R.	Readings in one direction		Total (*a*)
		C.S.R.	Corrected C.S.R.	
1.				
2.				
3.				

Mean diameter = ...

Readings in mutually perpendicular directions				Diameter $= \frac{a+b}{2}$
M.S.R.	C.S.R.	Corrected C.S.R.	Total (*b*)	

Mean radius $r =$ ....

Calculations

1. $$\rho = \frac{X}{l_2 - l_1}$$ (calculate for each set of Table. (A)).

2. $$X - Y = \rho(l_2 - l_1)$$
and $$Y = X - \rho(l_2 - l_1)$$
Knowing resistance of the wire Y

Calculate the specific resistance $\sigma = \frac{Y\pi r^2}{l}$

Precautions

1. The connections must be neat and tight. Small and thick wires should be used.
2. To avoid contact resistance the dial type of resistance box should be used.
3. Pressed jockey should not be moved in wire. It should be pressed gently and momentarily to avoid heating.
4. The difference between X and Y should not be more than the resistance of bridge wire, otherwise balance point will not be obtained.
5. While determining ρ the balance should be obtained near the end of wire while for difference of resistance it should be obtained near the middle of the wire.
6. Bridge is most sensitive when P, Q, X and Y are of the same order.

Questions with Answers

1. Define resistance?

Ans. $R = V/I$, where V is the potential difference applied across the ends of the conductor and I is the current in amps flowing through the conductor.

2. How the resistance varies with temperature?

Ans. Resistance of metals increases with increase of temperature, $R_t = R_0(1 + \alpha t + \beta t^2)$. However, the resistance of semiconductors decrease with increase of temperature. There are certain metals and alloys whose resistance vanishes suddenly in the vicinity of absolute zero. These are called superconductors.

3. What is specific resistance?

Ans. Specific resistance of a material of a wire is the resistance of one centimeter length of the wire whose cross-sectional area is one square centimeter

$$K = \frac{\pi r^2 R}{l} \text{ ohm - cm}$$

where r is the radius, l is the length of the experimental wire and R is the resistance of wire.

4. If the radius of the wire is doubled will its specific resistance change?

Ans. We may note that specific resistance is independent of length or radius of the wire and depends only on the material of the wire.

5. How in a resistance box, resistance coil are placed?

Ans. Insulated wires of maganin of suitable length are doubled on itself and then it is wound on a non-conducting material. The free ends of these wires are then soldered to the piece of the brass at the top.

6. What is the purpose of doubling the resistance wire prior to winding over the non-conducting material, i.e. bobbin?

Ans. In order to avoid the inductive effects. Since the direction of current in one wire is opposite to that of the current in the other wire and hence the total flux is always zero.

7. What is the principle of Wheat stone bridge?

Ans. When the resistance in the four arms P, Q, R and S of Wheat stone bridge are such that there is no deflection in the galvanometer, then we have

$$\frac{P}{Q} = \frac{R}{S}$$

8. Explain the principle of Carey Foster's bridge.

Ans. When the resistance connected in the outer gaps are interchanged, the position of null point changes on the bridge wire. The difference in the value of resistance is equal to the resistance of the differences of bridge wire, i.e. $\rho(l_2 - l_1)$.

9. When Carey Foster's bridge is most sensitive?

Ans. When P, Q, R and S are equal or of the same order, Carey Foster's bridge is most sensitive.

10. What is the maximum difference between the two resistances which can be measured by Carey Foster's bridge?

Ans. Equal to total resistance of the Carey Foster's bridge wire.

11. What is the minimum difference in resistances that can be measured?

Ans. The resistance of 1 mm length of the Carey Foster's bridge wire.

12. Why do we keep the resistance of inner ratio arms equal?

Ans. This increases the sensitivity and accuracy of Carey Foster's bridge.

13. What do you understand by end resistances?

Ans. The resistance of the copper strips fitted at the end of Carey Foster's bridge and that of solder are not taken into consideration. These are termed as end resistances.

14. Why do we interchange the position of resistance box and unknown resistance in the outer gaps?

Ans. See ans. 8.

EXPERIMENT 25

Object

To determine the waveform and frequency of an electrically maintained tuning fork by forming Lissajous figures on CRO.

Apparatus

Cathode ray oscilloscope, Electrically maintained tuning fork, Standard audio frequency signal generator.

Description of C.R.O.

Cathode ray tube is an electronic device that produces visual representation of electrical effects. It consists of the following main constituents.

1. **Electron Gun:** The main parts of electron gun are heater, cathode, control grid, focussing anode and accelerating anode. This assembly generates and controls the electrons in a narrow beam at high velocity.
2. **Deflecting Device:** This system deflects the electron beam either electrostatically or magnetically in accordance with the phenomenon (voltage or current wavefrom) to be displayed. The electrostatic deflection is accomplished through the use of vertical (*YY*) and horizontal (*XX*) deflection plates. These plates deflect the beam vertically or horizontally as the case may be. For the deflection of beam magnetically, the magneti field is produced by coils ouside the tube.
3. **Time Base Circuit:** The horizontal deflecting plates XX are connected to a 'time base' voltage of known frequency of alteration. This voltage causes the spot of sweep across the screen (along the *X*-axis) at a uniform speed and then to quickly flash back to the starting point. The potential difference of which the waveform is to be investigated is applied between the plates *YY*.
4. **Fluorescent Screen:** Upon which the electron beam is passed to create a well-defined spot and which emits light when bombarded by electrons as this screen is coated with fluorescence matereial, viz. ZnS, $ZnSiO_4$, etc.

   Other main constituents are deflection voltage amplifiers, power supplies, synchronising selector, frequency selector and fine frequency knob, etc.

   Figure 51 shows the essential parts of a cathode ray tube using electrostatic focussing and deflection.

Working

If the alternating voltage is applied to the *XX* plates only the spot sweeps a horizontal line passing through *S* and if the voltage is applied to the *YY* plates the spot moves along a vertical line passing through *S*. If both the fields are applied, a waveform representing the voltage applied to *YY* plates is traced and by synchronising this voltage with the time base voltage, a fixed pattern of the waveform is obtained. In case two alternating potentials with simple harmonic variations are applied simultaneously between the plates *XX* and *YY*, the electronic beam traces Lissajous figures. A grid structure between cathode and anode controls the intensity of beam.

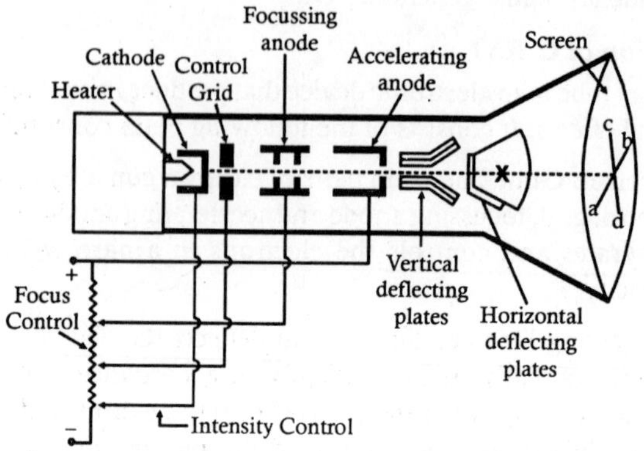

Fig. 51: Structure of electro-static deflection type cathode ray tube.

The extremely small inertia of the electrons enables the electron beam to follow the variations in electric and magnetic fields which deflect it with practically no time lag. Persistence of vision and presistance of fluorescence give the illusion of an image.

Theory

A cathode ray oscilloscope can be conveniently used to determine the frequency of an electrically maintained tuning fork, by applying the unknown frequency (voltage) to *Y-Y* deflection plates and a known frequency to *X-X* deflection plates from a standard signal generator. If the two frequencies have a constant ratio, simple Lissajou's figures as shown in Fig. 52 are obtained on the screen. By counting the tangential points on the sides of the rectangle enclosing the Lissajou's figure, the ratio of the two frequencies can be determined by the following relation.

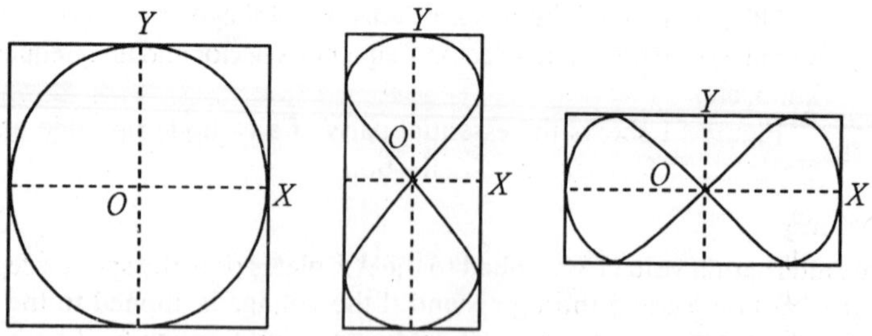

$\frac{f_v}{f_h} = \frac{1}{1}$ $\frac{f_v}{f_h} = \frac{1}{2}$ $\frac{f_v}{f_h} = \frac{2}{1}$

Fig. 52

$$\frac{f_v}{f_h} = \frac{\text{Number of tangential points on the horizontal axis}}{\text{Number of tangetial points on the vertical axis}}$$

If one frequency is known, the other can be calculated.

Procedure

Set up the electrical circuit as shown in the circuit diagram (Fig. 53).

The experiment is performed in the following two stages:

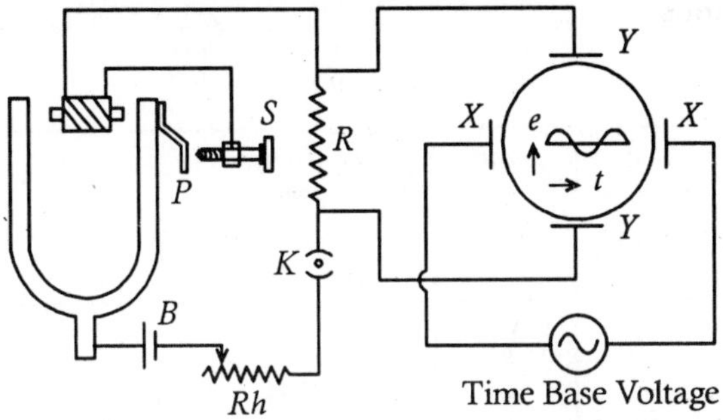

Fig. 53

(a) Determination of the Waveform: Switch on the power supply of *CRO* and obtain a sharp and fine point of the electron beam at the centre of the screen by adjusting the intensity and focussing control knobs of the *CRO.*

Make the connections as shown in Fig. 53 by introducing a suitable resistance R say 100 ohms. Now a voltage will be developed across the resistance R . This voltage is applied to the *Y-Y* plates of *CRO.*

Set the tuning fork into vibrations by means of the screw S till it touches with the spring P. A vertical line will be observed on the screen of *CRO.* Use Y-amplifier to keep this line within the scale.

Now apply the time base to *X-Y* plates of the *CRO* and synchronise it to obtain a steady pattern on the screen. The pattern obtained will be the waveform of electricity maintained tuning fork.

Trace the waveform on a tracing paper.

(b) Determination of Frequency: Connect a standard audio frequency generator to *X-X* plates and change the selector knob of *CRO.*

Adjust the frequency of the signal generator such that a circle or an ellipse of suitable size is obtained on the screen. Trace this circle (or ellipse) on a tracing paper and note down the frequency of the signal generator. The frequency of the signal generator will be equal to the frequency of the electrically maintained tuning fork.

Again adjust the frequency of the signal generator such that a figure of eight (8) is obtained on the screen. Trace this figure and note down the frequency of the signal generator. In this case the frequency of the electrically maintained tuning fork will be half of the frequency of signal generator.

Now, again adjust the frequency of the signal generator such that a figure of eight is obtained in horizontal direction (∞). Trace this figure and note down the frequency of the signal generator. In this case, the frequency of the tuning fork will be double the frequency of the signal generator.

Observations

TABLE 1

S.No.	*Horizontal input frequency*	*Shape of pattern*	*No. of tangential points on*		$\frac{f_v}{f_h}$	*Vertical frequency*
			Horizontal axis	*Vertical axis*		
1.	...	O	...	...	...	...
2.	...	8	...	...	...	...
3.	..	∞	...	...	...	...

Results

1. The waveform of the electrically maintained tuning fork is shown on the attached tracing paper.
2. The frequency of the electrically maintained tuninf fork = ... cycles/sec.

Questions with Answers

1. What is a cathode ray tube?

Ans. It is a vacuum tube used in the formation of *CRO*.

2. What is the main function of *CR* tube?

Ans. *CR* tube allows the direct observation of the behaviour of cathode rays (electrons) and provide a visual representations of voltage and current forms.

3. What are the main parts of a *CR* tube?

Ans. A *CR* tube consists of a highly evacuated funnel shaped glass envelope containing an electron-gun, a fluorescent screen and a deflection system.

EXPERIMENT 26

Object

To obtain hysteresis curve for a given ferromagnetic material on CRO using solenoid.

Materials Required

1. The specimen in the form of several wires of iron ≃ 1 mm diameter.

2. A solenoid wound on a glass tube: diameter $\simeq$1 cm to 2 cm length of winding 20 cm to 30 cm, two layers of No. 22 or 24 wire (enamelled or double cotton covered).
3. Secondary coil 10,000 turns of No. 46 or 48 enamelled wire, length <10 cm, diameter $\simeq$ 5 cm.
4. A 6-volt stepdown transformer (2 amp capacity).
5. An A.C. ammeter (2 amp range), or an A.C. voltmeter used across a known resistance.
6. Rheostat, about 6 Ω resistance.
7. Resistors: 1.0 Ω(two), carbon resistor 47 k Ω (one).
8. Capacitors: 2 μF (one) 1 μF (one)
9. Cathode ray oscillogarph.

Apparatus

The diamensions of the solenoid, the secondary, coil, etc. may be considered flexible. Similarly, the values of resistors and capacitors. The cathode ray oscillograph need not have a time base. A simple low-frequency oscillograph with (about) 8 cm screen is enough.

Instead of an anchor ring-type closed magnetic circuit, the solenoid arrangement is used to enable us to specimens at will, and, what is more important, to enable us to get a caliberation for the *B* value measurement. This choice, however, introduces some limitations due to the demagnetization effect of the open ends of the sample, etc.

Theory

To obtain the *B-H* curve for a sample experimentally, one has to apply to the sample a varying magnetic field *H* of an amplitude large enough to attain saturation and measure the corresponding variations of *B*. In the present arrangement, *H* is produced by 50 cycles A.C. $i = i_0 \cos\omega t$ in the solenoid, giving

$$H = \frac{4\pi ni}{10}\cos\omega t = \frac{4\pi ni\sqrt{2}\, i_{rms}}{10}\cos\omega t \tag{1}$$

where i_{rms} is the current in amp measured by the ammeter, and n is the number of turns per cm in the solenoid. The voltage across a 1.0 Ω resistance R_1 (Fig. 54) is fed to x plates of the C.R.O. The x deflection is, therefore; propotional to *H*.

The e.m.f. across the secondary coil is a measure of (dB/dt) not of *B*. Hence, an integrating circuit is used. In Fig. 54, the 47-kΩ resistance (*R*) and the 2 μF condenser present this circuit. The voltage across the condenser at any instant is given by

$$V = \frac{Q}{C} = \int_0^t \frac{i_s dt}{C} = \int_0^t \frac{Edt}{RC}\ \alpha\ \frac{B}{RC} \tag{2}$$

where E is the induced e.m.f. and i_S the current through the secondary. Eq. (2) holds provided that R is much greater than the reactance of the circuit. Thus, V is proportional to B. When applied to Y plates of the C.R.O. it would produce a Y deflection proportional to B.

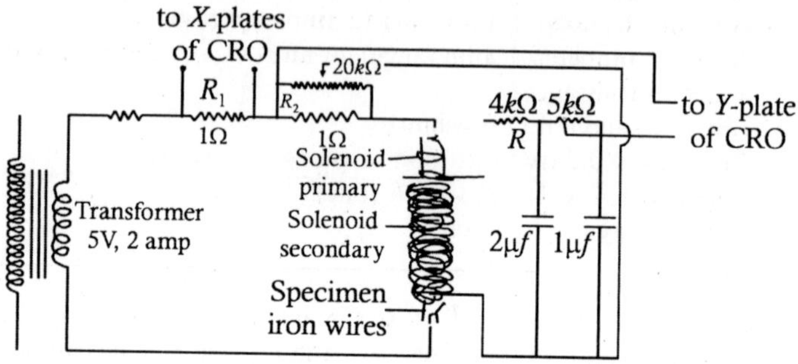

Fig. 54: The circuit diagram. (Read R = 47 kΩ, 4kΩ)

In practice, however, R cannot be made very large compared with the reactance. Therefore, there is a phase difference between i_s and E and, hence, between V and B. This is corrected by the 1 μF condenser and the variable resistance (5 kΩ) shown in Fig. 54.

The coupling between the secondary and the primary is partly through iron and partly through air. The latter makes a contribution to B which is in phase with H (hence with the primary current i), while the former (which we desire to measure) is in phase with (di/dt). To neutralize the air contribution, the potential drop over a resistance in the primary is suitably added with the voltage to the applied to Y plates. The special arrangement shown as R_2 plus a 20 kΩ potentiometer across it acts as this correcting device. Many of these details will be best understood while actually observing the pattern on the C.R.O.

Procedure

1. Complete the circuit connections as shown in Fig. 54 and switch on. Some pattern will appear on the C.R.O. screen.
2. Adjust the x and y amplifiers of the C.R.O. to get a pattern within the screen.
3. A shape like that shown in Fig. 55 (*e*) is the correct shape of a *B-H* curve. But shapes like those shown in Fig. 55 (*a*) to (*c*) may appear. (The secondary leads to the C.R.O. may have to be interchanged to bring the curve in the proper quadrants).
4. Adjust the 5 kΩ rheostat (Fig. 54) in the phase connecting circuit to do away with the loops or flats at the tips of the pattern. This would

give a curve like that shown in Fig. 55 (*c*) with sharp tips. But these ends may not be parallel to the *H*-axis (horizontal). This is due to the contribution of the flux in the air core.

5. Then, adjust the 20 kΩ potential divider in the primary circuit to make the ends of the curve horizontal. (The connections at R_2 may have to be reversed.)

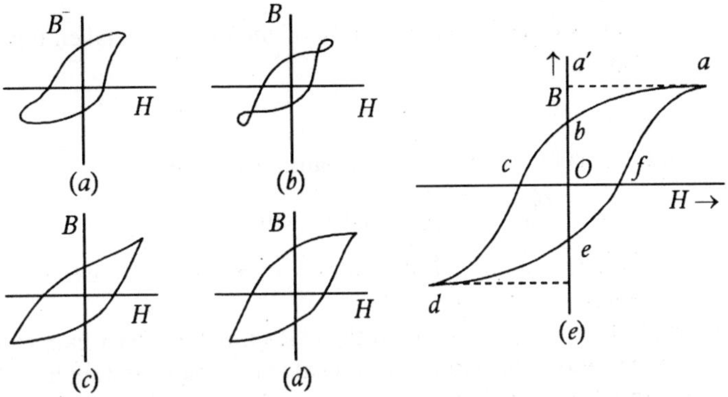

Fig. 55: Possible figures seen on the CRO, (*a*) to (*d*) stages of improvement (*e*) the appropriate B.H. curve

6. If the tips of the curve still do not show a sharp overlapping region [Fig. 55 (*d*)], this means that magnetic saturation has not been achieved. Increase the current by using the rheostat in the primary.
7. You now have the *B-H* curve [Fig. 55 (*e*)]. You can change its *H* or *B* scale by adjusting the *x* and *y* amplifications of the C.R.O. In this curve the quantities are: *Oa'* the saturation magnetization B; *Ob* the residual magnetization (retentivity); and *Oc* the coercive field. But without calibration, you do not get their magnitudes. You can only trace the curve on a tracing paper. The ratios (Ob/Oa') and ($Oc/a'a$) are obtained even without caliberation. Interpret them.

Caliberating the *H* and *B* Axes of the *B-H* Curve Obtained on the C.R.O.

1. Measure the maximum *x* deflection L_x (=2*a'a*) on the observed *B-H* curve [Fig. 55 (*e*)]. and the current *i* (rms value is read by the meter). Then, the caliberation constant *c* for the *H* axis is given by

$$c = \frac{2H_{\max}}{L_x} = \frac{8\pi ni\sqrt{2}}{10L_x} \text{ oersted / cm} \tag{3}$$

where *n* is the number of turns/cm in the primary, *i* is in amp and L_x in cm.

2. Measure the maximum y deflection $L_y (= 2Od')$ on the observed B-H curve [Fig. 55 (*e*)]. Now, set the 20 kΩ potential divider at zero and remove the iron core. The y deflection would fall to L'_y, which corresponds to the magnetic flux ϕ with air as the medium, and this is known to be $H\pi r^2$, where r is mean radius of the primary. Thus,

$$\frac{\phi_{max}}{(\phi_{air})_{max}} = L_y/L'_y \rightarrow \phi_{max} = H_{max}\pi r^2 . L_y/L'_y \tag{4}$$

If s is the area of cross-section of the iron wires taken together, we have

$$2B_{max} = \frac{2\phi_{max}}{s} = AL_y \tag{5}$$

where A is the caliberation constant for B, given by

$$A = \frac{2\phi_{max}}{sL_y} = \frac{2H_{max}}{sL'_y} \text{ gauss / cm} \tag{6}$$

But determining L'_y is difficult because $L'_y << L_y$ in practice. The procedure is as follows.

3. With the hysteresis curve on the C.R.O. reduce the x gain to zero (so that only a line along the vertical axis is seen), and set the 20 kΩ potential divider (Fig. 53) at zero (so that it offers no resistance in the secondary circuit). Measure L_y.
4. Now, move the magnetic material slowly out of the solenoid till the length of the vertical line on the screen is reduced to $(L_y/2)$; increase the y gain to restore this length to L_y; move the magnetic material further out to reduce this length to $(L_y/2)$; increase the y gain further to restore this length to L_y, and so on.
5. Continue to proceed as above till the material is nearly out of the solenoid. Let us say it takes p steps. In the final stage, take the materieal totally out and measure the length h on the screen. This gives $L'_y = h \times 2^{-p}$.
6. Now, compute the caliberation constant A[Eq. (6)] for the B axis.

EXPERIMENT 27

Object

To design and study amplifying characteristics of a single state RC coupled amplifier.

(A) Single Stage Amplifier Using a Triode

Aim

To design a single stage amplifier using a triode and study its frequency response and study the variation of gain with load resistance.

Apparatus and Components

A 6J5 valve, power unit, an audio-frequency oscillator of range 0-50,000 Hz, base board, resistors, capacitors, etc.

Procedure

The connections are made as shown in Fig. 56 by effective soldering. Here the triode valve is biased for class *A* operation, since in this case distortion produced is minimum. The input is fed in the grid through the coupling capacitor 0.047 μF and the grid leak resistance 470 K. The resistance 1 K and the electrolytic capacitor 32μF, 25 V ensures proper bias conditions.

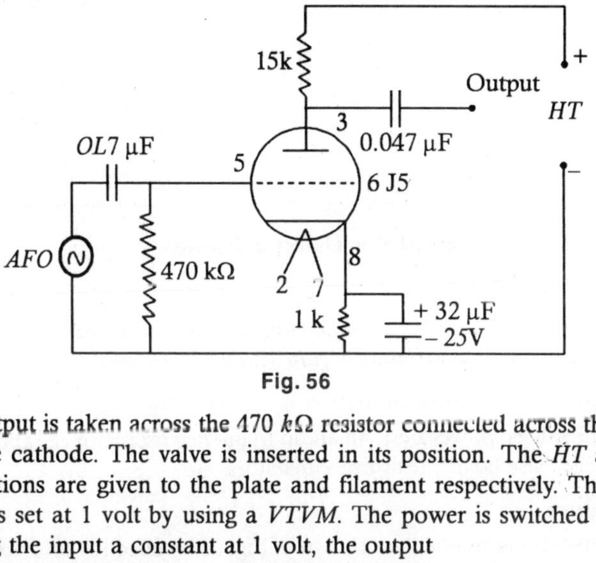

Fig. 56

The output is taken across the 470 $k\Omega$ resistor connected across the plate and the cathode. The valve is inserted in its position. The *HT* and *LT* connections are given to the plate and filament respectively. The input signal is set at 1 volt by using a *VTVM*. The power is switched on and keeping the input a constant at 1 volt, the output voltage is measured using a *VTVM* for different input frequencies ranging from 50 Hz to 50 kHz.

A graph is drawn taking logarithm of the frequency along the *X*-axis and the gain along the *Y*-axis. The graph is as shown in Fig. 57. The band width of the amplifier is the difference in frequency $(f_2 - f_1)$ between the half power points where the gain is $A/\sqrt{2}$. The experiment is repeated for different values of $R_L = 33\,k\Omega$ and $47\,k\Omega$.

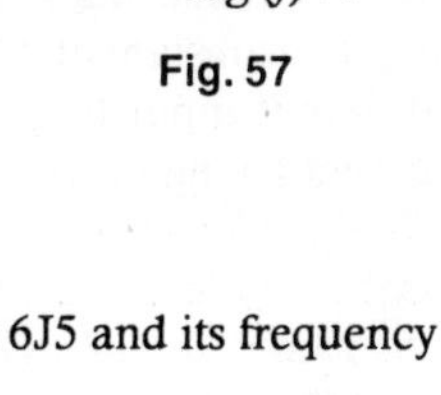

Fig. 57

Result

A single stage amplifier is constructed using a triode 6J5 and its frequency response is studied.

TABLE 1

Frequency fHz	*Output* *volt*	*Gain* (A) *Output voltage* / *Input voltage*	log f
50			
100			
300			
500			
600			
1000			
...			
...			
...			
20000			
30000			
40000			
50000			

(B) R-C Coupled Amplifier Using a Triode

Aim

To design an R-C coupled amplifier usng triodes and to study its frequency response and study the variation of gain with load resistance.

Apparatus and Components

Two 6J5 valves, power unit, an audio frequency oscillator of range 50 Hz to 50 kHz, base board, resistors, capacitors, etc.

Procedure

The connections are made as shown in Fig. 58 by effective soldering. Here the output of the first stage amplifier valve is fed as the input across the grid and the cathode of the second stage amplifier valve. Since the triode amplifier has a large input impedance, it can be coupled on to the next stage without loss of power.

The valves are inserted in their positions. The *HT* and *LT* connections are given to the plate and filament respectively. The input signal is set at 0.1 volt by using a *VTVM*. The power is switched on and keeping the input a constant at 0.1 volt, the output voltage is measured by a *VTVM* for different input frequencies ranging from 50 Hz to 50 kHz. A graph is drawn taking logarithm of the frequency along *X*-axis and the gain along *Y*-axis. The graph is as shown in Fig. 59. The band width of the amplifier is the difference in frequency $(f_2 - f_1)$ between the half power points, where the gain is $A/\sqrt{2}$. The experiment is repeated for different values of R_L = 33 $k\Omega$ and 47 $k\Omega$.

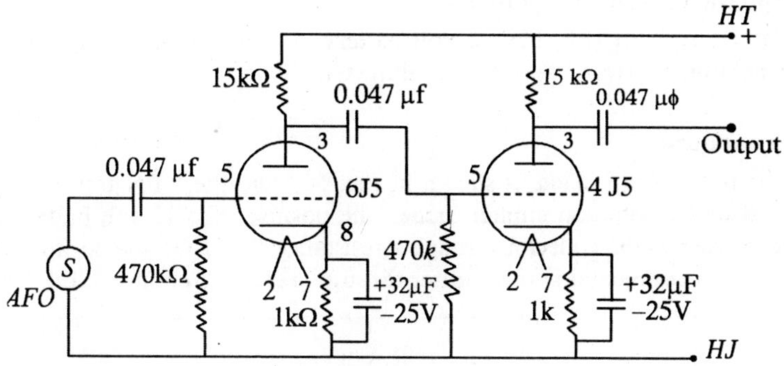

Fig. 58

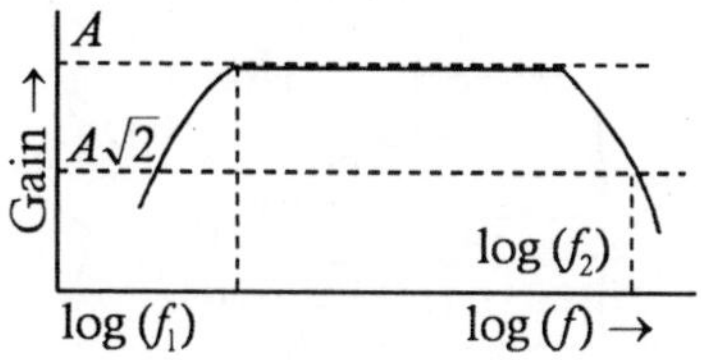

Fig. 59

Result

An *R-C* coupled amplifier is constructed using two triodes (6J5) and its frequency reponse is studied.

TABLE 2

Frequency (f) Hz	*Output* Volt	*Gain* $A = \frac{V}{0.1}$	*log (f)*
50			
100			
1000			
...			
20000			
50000			

(C) R-C Coupled Transistor Amplifier (Single Stage)

Aim

To construct a single stage R.C. coupled transistor amplifier and to find the voltage gain.

Apparatus and Components

NPN transistor (BC 148), 2 volt battery, 12 volt battery, A.F. signal generator, resistors, capacitors, multimeter, etc.

Procedure

Connections are made as shown in Fig. 60 Here the NPN transistor is used under common emitter mode. The positive of a 12 volt battery is connected to the collector through a resistance R_3 ($2k\,\Omega$), the negative of which is connected is the emitter through a R_E—C_E combination. The

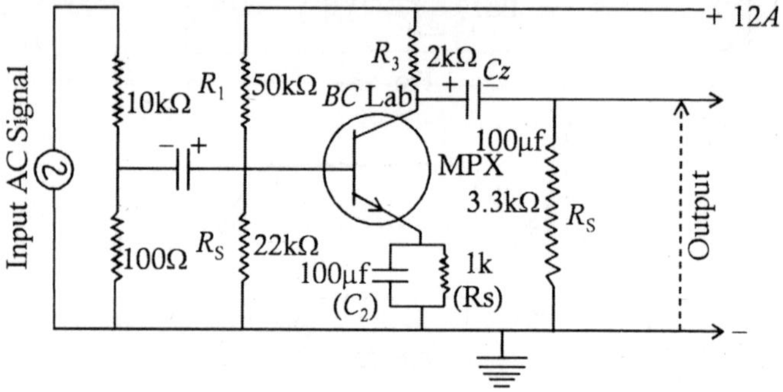

Fig. 60

A.F. signal from a signal generator of a given frequency is applied between the base and the emitter through a coupling capacitor C_1(100 µF) and resistors R_1 (50 k Ω) and R_2 (22 kΩ) which serve as potential divider. This arrangement provides proper bias to the emitter and base.

The input signal voltage and the output voltage across R_L(3.3 kΩ) are measured using a multimeter.

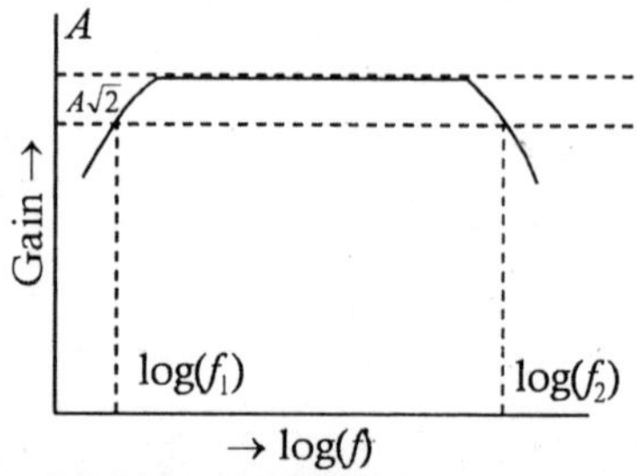

Fig. 61

The voltage gain $= \dfrac{\text{Output voltage}}{\text{Input voltage}}$. The gain is measured for different frequencies and frequency reponse curve is drawn.

Result

A single stage R.C. coupled transistor amplifier is constructed. Its voltage gain is measured at different frequencies and the frequency response curve is drawn.

Questions with Answers

1. What is an amplifier?

Ans. It is a device which increases the value of A.C. signal applied to its input.

2. How are amplifiers classified?

Ans. We classify amplifiers according to operating conditions under which the valve transistor works. They are classified as class *A*, class *B*, class *AB* and class *C*.

3. What is a cascade amplifier?

Ans. In order to obtain large amplification as compared to a single stage amplifier a number of amplifiers are used. Each is called a stage. In a cascade, the output of one stage is fed as input as next stage and so on. This arrangement of amplifiers is called cascading of amplifiers.

4. What is voltage gain of an amplifier?

Ans. The ratio of output voltage to input voltage.

5. What is phase relationship between input signal and output signal?

Ans. The phase difference between them is 180° or π, i.e., they are in opposite phase.

6. Mention most commonly used coupling methods?

Ans. (*i*) RC coupling (*ii*) LC coupling (*iii*) Transformer coupling (*iv*) Direct coupling.

7. How does a transistor work as an amplifier?

Ans. For amplification, energy is supplied by the bias voltage. A small change in the input current in a transistor circuit causes a large variation in the output current. This is why a transistor works as an amplifier.

8. Why does the amplification falls off at very low and very high frequencies?

Ans. The fall of amplification at very low frequencies is due to the fact that the reactance of the coupling capacitor is very high at these frequencies, whereas the fall at very high frequencies is due to various capacitances (input, output, stray capacitances) which shunt the coupling and base leak resistance.

9. What do you understand by half power frequencies?

Ans. The frequencies at which the voltage gain of the amplifier falls to $1/\sqrt{2}$ times its mid-frequency gain, are termed as half power frequencies.

10. Define bandwidth.

Ans. $\beta = f_n - f_i$, i.e., the difference of upper and lower half power frequencies is called bandwidth.

11. What are the main advantages of using transistor in a common emitter configuration as an amplifier?

Ans. In *CE* configuration, the current gain, the voltage gain and the power gain are high. Due to being of some order (kilo ohm) input and output impedences, power transmission takes place easily. We may note that input and output impedences are purely resistive.

EXPERIMENT 28

Object

To draw the plateau characteristic of a Geiger Muller counter using a radioactive source.

Apparatus

Geiger Muller tube, High Tension Battery with regulator, Radioactive source, A resistor, A capacitor, Pulse Amplifier, Electronic scalar circuit, the Chemical Recorder.

Description of Apparatus and Theory

A Geiger Muller counter is simply is a gas filled nuclear detector based on the ionisation of a gas by elementary particles and operating in a certain specified region of applied potential difference, known as *Geiger muller region*.

A gas filled detector usually consists of a gas filled chamber (whose conducting walls act as cathode) and a control electrode (working as anode). A potential difference is applied between cathode and anode by a high tension supply through an external resistor *R*. When a high electric radiation or charged particle enters through such a chamber, the ionisation of the gas starts resulting in the formation of ion-pairs. In the absence of electric field, the ion pairs thus created will just recombine. It has been proved that formation of ion-pair in air about 35 eV energy is required. If the incident radiation of charged particle loses 1 MeV energy in the chamber, it will form nearly 2.86×10^4 ion pairs.

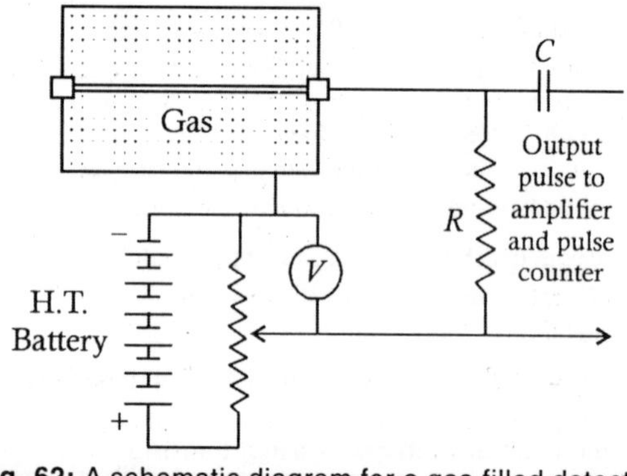

Fig. 62: A schematic diagram for a gas filled detector

As the potential difference is applied, the ion pairs are collected by the respective electrodes; thus giving rise to a current pulse. This current pulse passing through resistor *R*, gives rise to electrical pulse signal which can be amplified and then detected by a pulse counter.

A plot of output pulse height at anode (or relative number of ions collected by the anode) will depend on (*i*) external voltage applied and (*ii*) initial ionizing agent-that is whether the pulse has been initiated by the passage of a particle of high or low ionizing power. Figure 63, represents the typical plot of log of pulse height versus applied potential difference *V* for two different types of events in the counter. One event is the passage of a low ionizing β-particle through the gas while the other is the passage of highly ionising α-particle. The nature of the graph may be understood by dividing it into four regions:

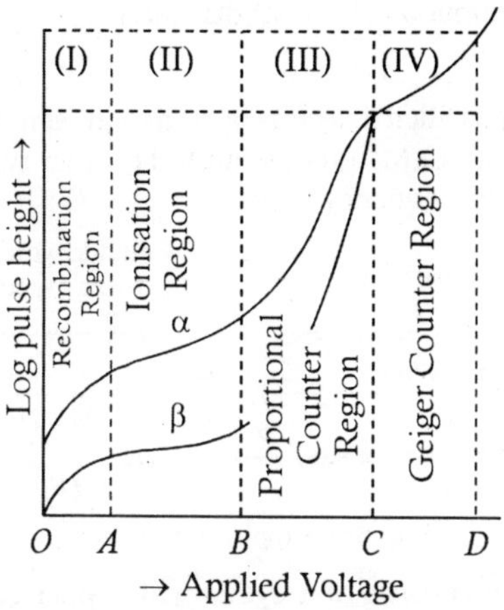

Fig. 63: General characteristics of a gas field detector. Curve α represents the ionisation due to highly ionising particles, while β represents the ionisation due to low ionising particles.

1. Initially when the applied voltage is low, the electric field is so low that the recombination of positive and negative ions takes place; as a result all the ion-pairs formed are not detected. As the electric field is increased the ions move faster leaving less time for possible recombination and thus a relatively larger number of ions reach the electrode. The region *I* or *OA* is therefore called the *Recombination* region.
2. As the voltage is increased further, region II is reached where the loss of ions through recombination is almost negligible, this region is called the *Ionisation Chamber region.*
3. In region III, the voltage is sufficiently high so that the electrons liberated by the primary ionising event get sufficient energy to cause secondary ionisation. The secondary ionisation results ion an increase of collected charge. In this region the pulse height depends on the initial ionising event.

   Greater is the ionising power of initial particle, greater is the pulse height. Therefore, this region is called the *proportional region.* As the applied voltage is increased this proportionality breaks down, the upper end of this region is called the region of *limited proportionality.*
4. Beyond the proportionality region IV is reached where the pulse height is completely independent of the initial ionisation and all

particles produce pulses of same height irrespective of their energy and primary ionisation. This region is called *Geiger-Muller* region.

If the voltage is increased beyond region IV, there will be an onset of continuous electrical discharge.

G.M. Tube

G.M. counters are gas filled detectors operating in region IV. A considerable variety of designs of G.M. tube is built. The two types of G.M. tubes shown in Fig. 64 are frequency used.

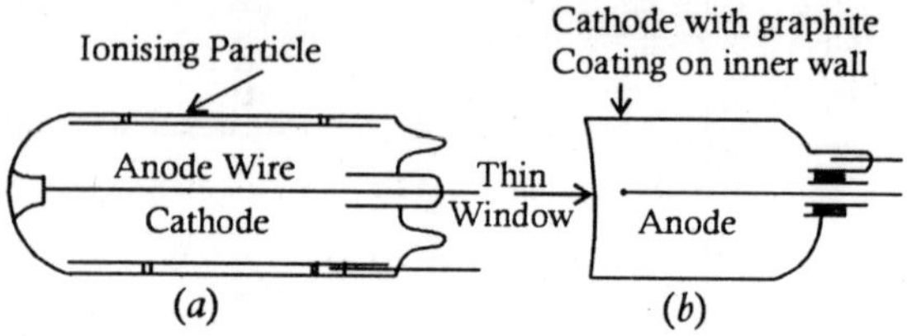

Fig. 64: Simple G.M.Tubes

The first of these consists of a cylindrical copper cathode with a thin coaxial tungsten wire anode supported inside a cylindrical glass envelope. The elementary ionising particle enters the tube through its glass walls.

The second type is an end-window counter. The window can be thin mica sheet or in some designs a hemisphere of thin glass. These thin windows are required for counting less penetrating radiations, e.g. α-particles and less energetic β-particles which are unable to penetrate the glass wall of conventional counter. The anode is a tungusten wire and the cathode is in the form of a graphic coating on inside wall of glass envelope.

Quenching Process

When the G.M. tube operates in Geiger region, the secondary electrons increase the current pulse by further ionisation of gas molecules. The object of the counter is to produce a single pulse due to entry of a single particle. The tube should not then give any succeeding spurious pulses but should recover as quickly as possible to be in the state to record the next entering partical. Therefore, it is desirable to probability the production spurious pulses following the main required pulse due to a single particle entry. The process of prohibiting, the undesirable secondary pulses is called the quenching. There are two methods of quenching known as internal and external quenching.

Internal Quenching

In this method a suitable polyatomic vapour (known as *quenching gas*) such as ethyl, alcohol or hydrogen gas is introduced along with the innert gas filled in counter tube. (Ratio 90% argon, 10% ethyl alcohol by weight).

The quenching gas must have the ionisation potential lower than that of inert gas. The quenching action may be understod as follows:

The ionisation potential of alcohol is 113.3 V and that of argon is 15.7 V. As a result the inert gas ions neutralised by transferring their charge to alcohol ions during their long path towards cathode. The alcohol ions produced, capture electron from the cathode and are neutralised. Hence there is no multiple pulsing and the discharge is quenched soon (fraction of a milli-second) after the initial ionisation. The alcohol ions dissociate after being neutralised and after 10^8 counts the amount of alcohol in the counter is exhausted resulting in a poor characteristics of G.M. tube. The cathode is coated with graphite to avoid chemical attack.

External Quenching

In this method a large series resistance is used in the counting circuit. Due to secondary emission of electrons a large current pulse is produced. This current produces a large voltage drop across R, thus lowering the potential difference between the electrodes of the counter, so that further gas ionisation is avoided. This technique is not usually employed as it requires large recovery time (known as dead time of counter) for the tube to operate for the next count.

Method

(*i*) The connections are made as shown in Fig. 65

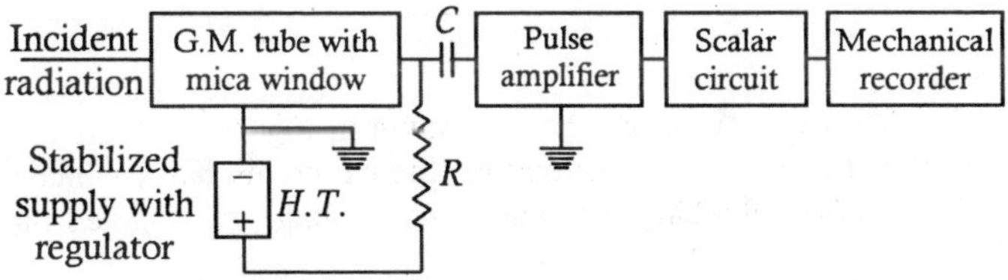

Fig. 65: Typical G. M. counter circuit

For an alcohol quenched circuit the operating voltage is kept 1000 to 1200 V; for bromide quenched circuit the operating voltage is kept 200 to 300 V. R is a resistance of 2 to 5 mega-ohm connected in series with counter and H.T. supply. The basic scalar circuit unit is essentially a twin-triode valve arrangement able to give a single output for two input pulses. The use of n such "scales of two" in cascade then enables the number of pulses produced by the tube to be reduced in the ratio 2^n : For example if $n = 10$; then $2^{10} = 1024$ input pulses produce only I output pulse. The output pulses are finally fed to mechanical recorder (*e.g.* telephone meter) which records the number of output pulses in a known time.

(*ii*) The particle (e.g. α or β particles) are made to enter to G.M. tube at a constant rate. The operating voltage across tube is varied and for each voltage the number of pulses recorded per minute is noted.

A graph is plotted for the pulses recorded per minute against the applied voltage; a curve like Fig. 66 is obtained.

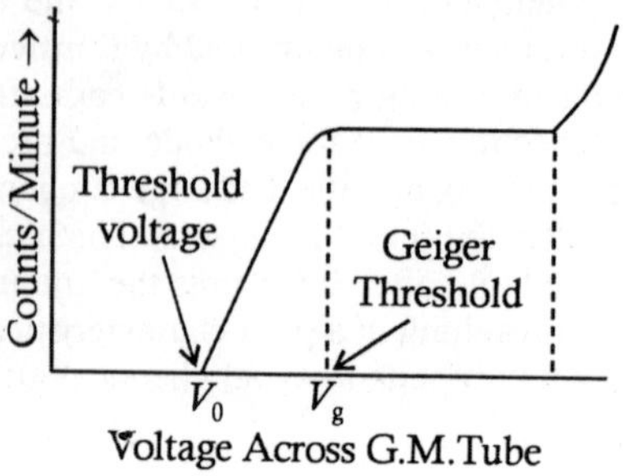

Fig. 66

Observations

S.N.	Operating voltage (volts)	Counts/min.
1.	... ...	... ...
2.	... ...	... ...
3.	... ...	... ...
4.	... ...	... ...
5.	... ...	... ...

Result

The characteristics of G.M. tube are shown in Fig. 66. The curve shows that for the applied voltage less than V_0, there is no discharge which means that there is no secondary ionisation. The voltage V_0 is called *threshold voltage*.

As voltage across G.M. tube is increased continuously a stage is reached when the pulse size is constant. This voltage is called *Geiger thereshold voltage*. Beyond this threshold further increase of voltage causes little increase of the number of counts. This gives *plateau region*, which is over 200 volts for a good tube. For G.M. counter the operating voltage is chosen to be on the *plateau*. The operating voltage minus the threshold voltage is *over-voltage* for the tube. The pulse size increases in proportion to the over voltage. Excessive over-voltage beyond the plateau region will lead to a continuous discharge.

G.M. counter tubes have a very limited life due to decomposition of organic vapour. Towards the end of their usual life the thereshold voltage is increased and the flatness of the plateau region is greatly reduced. These counters have a very low intrinsic efficiency for detection of γ-radiations.

G.M. counters have been very useful in detecting nuclear particles as well as measuring intensities of cosmic rays and recording cosmic ray events.

EXPERIMENT 29

Object

To study the variation for thermo e.m.f. with temperature for a thermocouple by means of a potentiometer.

Apparatus

Potentiometer, P.O, box , an accumulator, a standard cadmium cell (a freshly prepared daniell cell may also be used), two resistance boxes (1000 Ω), rheostat, sensitive galvanometer, two way key, Cu-Fe thermo couple, heating apparatus.

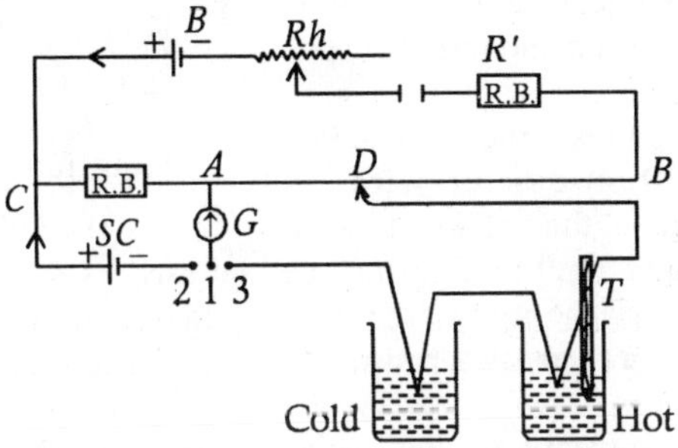

Fig. 67: Determination of thermo e.m.f.

Theory

Connections for the measurement of thermo e.m.f. are shown in Fig. 67. Since thermo e.m.f. is very small of the order of 0.13×10^{-4} volt per degree difference of temperature of two junctions of Cu-Fe couple. Hence, ordinary potentiometer device for measuring e.m.f. is not suitable. To measure such a small e.m.f. potential gradient on potentiometer wire should be very small. So also detecting device, i.e. galvanometer must also be very sensitive.

To obtain such a low potential gradient the arrangement is modified as follows:

Let a resistance R (above 1000 Ω) be connected in series with the potentiometer wire. The standard cell circuit is closed by putting the plug between terminals 1 and 2. The rheostat and resistance box in primary circuit are so adjusted that there is no deflection in galvanometer. Hence

Potential difference across R = E.M.F. of standard cell = E (say)

If I is current in potentiometer wire, then

$$E = IR \tag{1}$$

Now the standard cell circuit is broken and thermo-couple is connected by putting plug between 1 and 3. If l be the length of potentiometer wire across which the thermo e.m.f. 'e' balances, we have,

$$e = I \times (\text{resistance of } l \text{ cm of potentiometer wire}) = I\rho l \tag{2}$$

where ρ is resistance per unit length of potentiometer wire.

From Eqs. (1) and (2), we have

$$\frac{e}{E} = \frac{\rho l}{R} \quad \text{or} \quad e = \frac{\rho E l}{R} \tag{3}$$

which gives thermo e.m.f. in terms of measurable quantities.

A graph between thermo e.m.f. and temperature of hot junction above that of cold junction of thermo couple can be plotted (Fig. 68).

Procedure

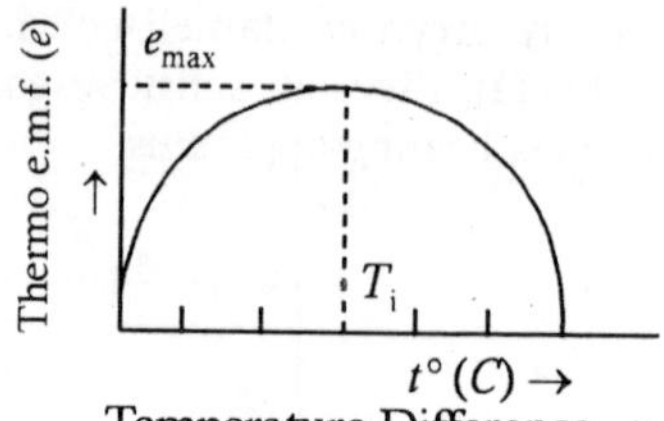

Fig. 68

1. To start with, first find resistance of potentiometer wire using P.O. Box.
2. Note the total length of potentiometer wire and find resistance per unit length.
3. Put a high resistance R in series with potentiometer wire. It should be of the order of 1000 Ω (preferably 1080 Ω if Daniell cell be used as S.C. or 1018.3 Ω if cadmium cell be used for the purpose). In series with this arrangement a battery, a rheostat, a resistance box and a plug key are connected (Fig. 67).
4. Connect standard cell with polarities as shown through a two way key and galvanometer across resistance box R. The thermo couple (Copper-Iron) should be connected between galvanometer and Jockey.
5. Close standard cell circuit by closing terminals marked (1) and (2). Introduce a resistance in R' nearly equal to that of R and by varying R' and R_h get zero deflection in galvanometer. Open key K and disconnect terminals (1) and (2).
6. Heat the junction placed in sand bath to a temperature about 5° above the cold junction.
7. Connect cold junction of thermo-couple by means of two way key to galvanometer by plugging terminals (1) and (3).
8. Find by trial the position of well point wire and note balancing length.
9. Repeat the observations of balancing lengths for different higher temperature at interval of 5° each up to say 300° C.
10. Take off heating arrangement and allow hot junction to cool. Find balancing lengths during cooling also at the same temperatures as has been done during heating.
11. Find mean of the two balancing lengths at same temperature during heating and cooling and calculate e.m.f. using relation (3) for each temperature separately.
12. Find difference of temperature between hot and cold junctions for each of the thermo e.m.f. calculated above.

13. Plot a graph between temperature on abscissa and corresponding thermo e.m.f. on ordinate.

Observations

(*a*) Length of Potentiometer wire L = ... cm.

(*b*) Determination of resistance of potentiometer wire:

S.No.	*Resistances in Ratio arms*		*Known resistance*	*Direction of deflection in Galvano-meter*	*Resistance of Poten-tiometer wire* ρ (Ω)
	P (Ω)	Q (Ω)	R (Ω)		
1.					
2.					
3.					
4.					

(*c*) Resistance per unit length of potentiometer wire $\rho = \frac{p}{L}$ = ohms/cm.

(*d*) Determination of thermo e.m.f.

(*i*) e.m.f. of standard cell E = ... volt.

(*ii*) Resistance R = ... ohm.

(*iii*) Temperature of cold junction = ... °C

S.No.	*Temperature of hot Junction*		*Balancing Lengths* cm			*Termo* e.m.f.
	As shown by Ther-mometer °C	*Temp. above cold junction* °C	*During increasing temp.* l_1 cm	*During decreasing temp.* l_2 cm	Mean $\frac{l_1 + l_2}{2}$ cm	$e = \frac{\rho El}{R}$ *volts*
1.						
2.						

Calculations

For resistance of potentiometer wire $p = \frac{Q}{\rho} R\Omega$

For $\rho = \frac{p}{L}$ = ohms/cm = ...

For thermo e.m.f. $e = \frac{\rho El}{R}$ ... volts

Result

The graph showing variation of thermo e.m.f. with temperature difference of junctions of Cu- Fe couple is plotted.

Precautions

1. The accumulator should be freshly charged and must have high capacity so that during the experiment it may have steady e.m.f.
2. The junctions of the thermo couple must be perfectly welded. Its leads should be large.
3. The jockey should only be momentarily pressed. In no circumstances it should be moved pressing, otherwise uniformity of wire will change which is chief source of error in measurements.
4. Battery key should only be pressed while taking observations.
5. Galvanometer used for detecting purpose must be very sensitive. During initial stages of balance, it should be shunted, and for exact balance the shunt should be removed.
6. Since non-uniformity for long wire is chief source of error in the expereiment, a Cromtions potentiometer or Sethi Raj-jada pattern potentiometer should be used.
7. If Daniell cell be used for standardisation, it should be freshly prepared.

EXPERIMENT 30

Object

To verify Stefan's law by electrical method.

Apparatus

D.C. voltmeter (range 0-4 V), D.C. ammeter (0-1 A), battery (6 V), rheostat (100 Ω), a bulb of tungsten filament.(6 V, 6 W).

Formula

To total power E emitted for a body at a temperature T_0 can be expressed as

$$E = C\left(T^{\alpha} - T_0^{\alpha}\right) \tag{1}$$

where α has a power ~4, C is a constant which depends on the material and area of body.

Rewriting Eq. (1) as

$$E = CT^{\alpha}\left(1 - \frac{T_0^{\alpha}}{T^{\alpha}}\right)$$

For $T >> T_0$, one can write

$$E = CT^{\alpha}$$

or $$\log_{10}E = \alpha\log_{10}T + \log_{10}C$$

on plotting a graph between $\log P$ and $\log T$, one obtains a straight line and its slope gives α. This verifies Stefan's law.

We measure the following parameters.

(*i*) Power radiated $E = VI$
(*ii*) Temperature of the body = °K

One can achieve it by measuring the resistance of filament R_t and then using the following relation.

$$R_E = R_0(1 + \alpha t + \beta t^2)$$

where α and β are known coefficients. Obviously, one can find the temperature of the filament.

In order to calculate, R_0 we first measure the resistance R_g when filament starts glowing and temperature at this stage is about 800 K. Now, using R_g ($= R_{800}$) and making use of the computed factor R_{800}/R_{273}, one can obtain the value of $R_{800}(=R_0)$.

Method

Connections are made as shown in Fig. 69. Now, make adjustment, such that each time the bulb glows for increasing and decreasing current. Now, find the ratio V/I, which gives R_g. Obviously, this is the filament resistance at 800 K. From the graph R_t/R_0 vs T, one finds

$$\frac{R_{800}}{R_{273}} = 3.9 \quad \text{or} \quad \frac{R_{800}}{3.9} = R_{273}$$

or $$\frac{R_g}{3.9} = R_0$$

one can draw the graph R_t/R_o vs T from the data given in Table 2.

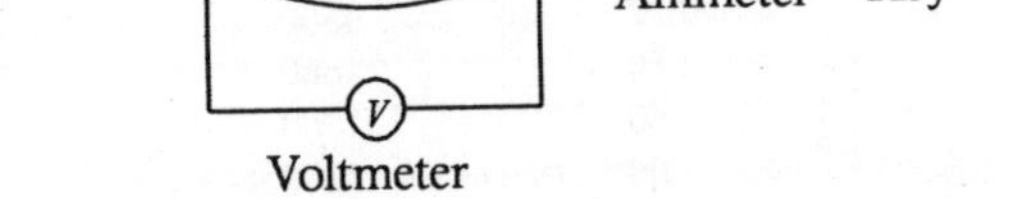

Fig. 69

Now, increase the current I from glow stage to values high enough to get dazzling white light and measure V across bulb each time. Obtain the value of power $E = VI$ and $R_t = V/I$, using these values of V and I. Now, from R_t using the value R_0 (or $R_g/3.9$) or R_{273}, one can obtain the temperature T of the filament of the bulb. Now, plot a graph between log$E$ and log$T$.

Observations

TABLE 1: OBSERVATIONS FOR R_g

S.No.	Current increasing			Current decreasing		
	V (volts)	I (amp)	$R_g = V/I\,(\Omega)$	V (volts)	I (amp)	$R_g = V/I$
1.						
2.						
3.						
4.						
5.						

Mean value of $R_g = \ldots\ \Omega$
From graph $R_{800}/R_{273} = 3.9$
Now, $R_o = R_g/3.9\ \Omega$

TABLE 2: MEASUREMENT OF POWER E

S.No.	V (volts)	I (amp)	$R_t = V/I\,(\Omega)$	R_t/R_o	$\log T$	$E = VI$ (watt)	$\log E$
1.							
2.							
3.							
4.							
5.							

Results

We obtain the graph of $\log E$ vs T as a straight line. The slope of line $\alpha = 4$. Obviously, $E = CT^{\alpha}$ i.e. Stefan's law is verified as fourth power law.

Temp. (°C)	R_t/R_0	Temp. (°C)	$R_t = R_0$
0	1.00	1100	7.60
100	1.53	1200	8.26
200	2.07	1300	8.90
300	2.13	1400	9.70
400	3.22	1500	10.43
500	3.80	1600	11.27
600	4.40	1700	11.42
700	5.00	1800	12.67
800	5.64	1900	13.50
900	6.37	2000	14.30
1000	6.94		

Questions with Answers

1. Define Stefan's law.

Ans. $E = \sigma T^4$ where $\sigma = 5.67 \times 10^{-8}$ watt/m^2-k^4 is Stefan's constant and E is the energy radiated per second by unit area of perfectly black body and T is the absolute temperature of the surface of the body.

2. Define emissive power of a body.

Ans. For a given wavelength at a given temperature, the emissive power of a body is the radiant energy emitted per second per unit surface area of the body per unit wavelength.

3. Define absorptive power.

Ans. Absorptive power of a surface is defined as the ratio of the radiant energy absorbed by the unit area of the surface in a given time to the total radiant energy incident on the same unit area of the surface in the same time.

4. What is a black body?

Ans. A black body absorbes all the incident radiations irrespective of frequency of radiations.

5. What is Kirchoff's law of radiation?

Ans. The ratio of the emissive power to the absorptive power of a particular wavelength at a given temperature is constant and it is equal to the emissive power of a perfectly black body at the same temperature for the same wavelength.

6. In this experiment, which is the radiating body?

Ans. Tungsten filament.

7. Explain the importance of Stefan's constant.

Ans. Stefan's constant is used in the determinition of the temperature of the heavenly bodies like sun.

8. What is the basic difference between the perfect body and non-perfect black body?

Ans. The emissivity of perfectly black body is unity, whereas the emissivity of non-perfectly black body is less then one.

EXPERIMENT 31

Object

To study the Hall effect and determine Hall coefficient, carrier density and mobility of a given semiconductor material.

Apparatus

Search coil, millivoltmeter, caliberated fluxmeter, battery, rectangular slab of a semiconductor crystal, ammeter, keys for circuit.

Formula

Magnetic field is applied along the Z-axis as shown in Fig. 70. Thickness of the crystal (d) is measured along Z-axis. Current (I_X) flows along X-axis and we measure the Hall voltage along Y-axis. Let permeability of the crystal is μ, then magnetic field, $B_z = \mu B$.

Now, Hall coefficient, $R_H = \frac{V_H}{I_X}\frac{d}{B_z} \times 10^4$. where V_H is Hall voltage (volts), ohm-m^3/W is current (ampere), d is crystal thickness (meter), B_z is magnetic field (Gauss).

Now, number of charge carriers per unit volume in the crystal.

$$n = \frac{1}{R_H e}\frac{me}{t^3}$$

(where n is electron concentration in n-type, semiconductor $e = 1.6 \times 10^{-19}$C)

Hall angle, $\phi = \frac{V_H}{V_x}\frac{l_x}{b}$ radian

where l_x and b are in meter. Now mobility,

$$m\mu = \frac{\phi}{B_z} \frac{\text{rad - m}^2}{\text{W}}$$

Method

First place the material in the magnetic field and then make the connections as shown in Fig. 70.

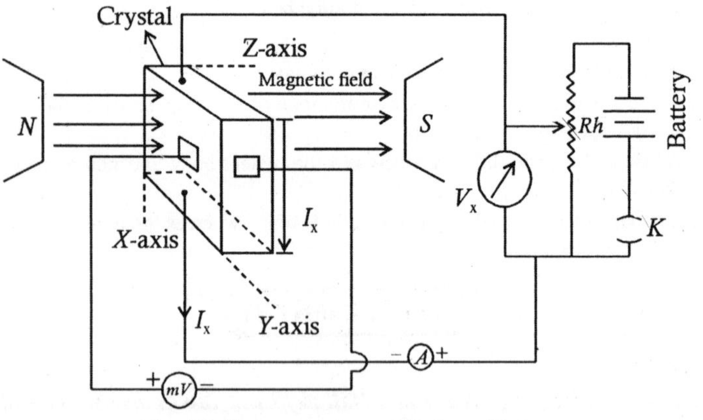

Fig. 70: Determination of Hall coefficient

Now, allow to pass the current I_x through the semiconductor crystal along X-axis. Measure the hall voltage V_H with the help of millivoltmeter (mV) and V_x by voltmeter (V) connected in the circuit. Now, change the value of I_x with the help of rheostat (Rh) and note down the corresponding values of V_x and V_H. Now plot a graph between V_H and I_x values. One obtains a straight line (Fig. 71). From this graph, one can calculate V_H/I_x, i.e. the slope of the line AB. With the help of a *gauss* meter or flux meter, measure magnetic field B and find actual field in the crystal $B_z = \mu B$.

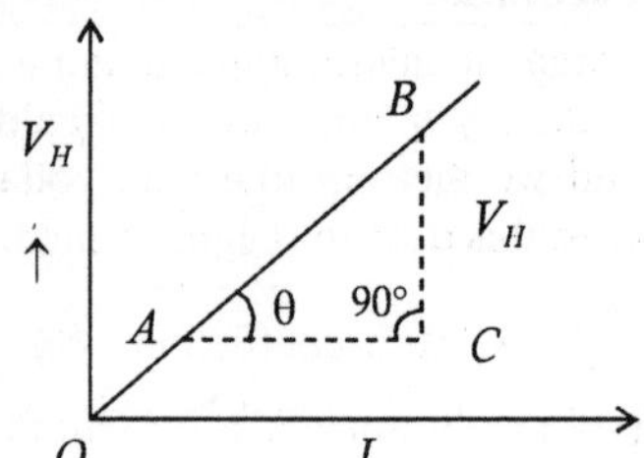

Fig. 71

Observations and Measurements

(*i*) Length of the crystal along *X*-axis, $l_x = \ldots$ m
(*ii*) Width of the crystal along *Z*-axis, $d = \ldots$ m
(*iii*) Width of the crystal along *Y*-axis, $b = \ldots$ m
(*iv*) Hall voltage = ...

Calculations

From graph between V_H and I_x (Fig. 71), we have

$$\tan\theta = \frac{V_H}{I_x} = \frac{BC}{AC}$$

Now, Hall coefficient

$$R_H = \tan\theta \frac{d}{B_z} \times 10^4 = \ldots \ \Omega \times \text{m}^2 / \text{W}$$

$$n = -\frac{1}{R_H e} = \ldots$$

Hall angle, $$\phi = \frac{V_H}{V_x}\frac{l_x}{b} = \ldots \text{ radian}$$

Mobility $$m_\mu = \frac{\phi}{B_z} = \ldots \text{rad - m}^2 / \text{W}$$

Result

Hall coefficient (R_H) = ... Ω-m^3/W
Number of charge carriers (n) = ...
Hall angle (ϕ) = ... rad
Mobility (m_μ) = ... $\frac{\text{rad} - \text{m}^2}{\text{W}}$

Questions with Answers

1. What do you understand by Half effect?

Ans. When we place a current carrying conductor in a magnetic field acting in a direction perpendicular to the direction of current, we find that a potential difference is developed at right angles to both current and magnetic field. We call this phenomenon Hall effect.

2. Define Hall coefficient.

Ans. We define Hall coefficient R_H numerically equal to Hall electric field ($E_H = V_H d$) induced in the specimen crystal by unit current, when placed perpendicular in a magnetic field of 1 W/m^2

3. What do you understand by mobility?

Ans. Mobility is the ratio of average drift velocity of the charge carriers to the applied electric field.

4. What is the unit of Hall coefficient?

Ans. Unit of R_H is Ohm-m^3/Weber.

5. What is the unit of mobility?

Ans. $m^2/(\text{volt-s})$

6. What do you understand by Hall angle?

Ans. $\phi = \frac{E_H}{E_x}$

when electron or hole is under the influence of simultaneous applied electric field (E_x) and Hall electric field (E_H) at right angles to each other, then the angle made with the x direction by the drift velocity V_x of the charge is called as Hall angle (ϕ).

7. When a transverse magnetic field is applied to a current carrying conductor, why Hall potential is developed?

Ans. When we place a current carrying conductor in a transverse magnetic field, the magnetic field exert a deflecting force $\boldsymbol{F} = e(\boldsymbol{v}_d \times \boldsymbol{B})$on the charge carriers in a direction perpendicular to both drift velocity $\boldsymbol{v}_d$ and applied field $\boldsymbol{B}$. Under the action of this force charges drift from one surface to another and as a result electric field is created due to excess negative charge at one surface and excess positive charge on opposite surface.

8. Explain the importance of Hall effect.

Ans. (*i*) One can calculate magnetic field from Hall potential difference.

(*ii*) One can decide whether a specimen is conductor, semiconductor or insulator.

(*iii*) One can measure the mobility of charge carriers.

(*iv*) With the help of the sign of the Hall potential difference, one can determine the nature of the charge carrier.

EXPERIMENT 32

Object

To measure the electrochemical equivalent of copper.

Apparatus

Copper voltmeter, Helmholtz galvanometer, copper plate, stop watch, power supply plate, variable resistance, four way key.

Formula

$$Z = \frac{m}{Kt}$$

where $\boldsymbol{m}$ is mass of copper deposited, $\boldsymbol{t}$ is time for which the current is passed and $\boldsymbol{k}$ is the reduction factor of Helmholtz galvanometer.

Procedure

Make the connections as shown in Fig. 72. Now, set the plane of coils of Helmholtz galvanometer in magnetic meridian. Now, adjust the current in the circuit with the help of variable resistance such that the deflection in compass box is 45° which should be kept constant in the 'On' position of the system. Now 'off' the circuit, and clean the cathode plate and weigh it and record it. Now, put the clean copper plate (cathode) again in copper voltmeter and after switch on the circuit allow to flow the current at least for 15 minutes in one direction and then change the direction of current in the Helmholtz galvanometer with the help of four way key. After 30 minutes time (t) interval, take the copper plate out from copper voltmeter. First dry it and then weigh the plate. Now, by subtracting the mass of cleaned copper plate from this mass, one can obtain the mass of copper deposited.

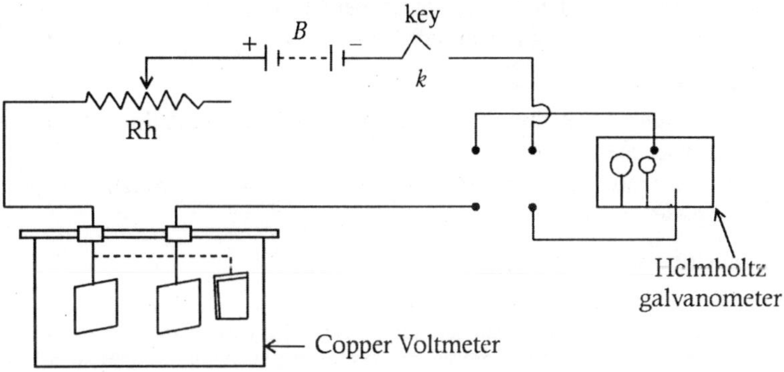

Fig. 72: Circuit diagram for the measurement of electrochemical equivalent of copper

Observations

(*i*) Mass of the copper plate (m) = ... gm

(*ii*) Mass of the copper plate + copper deposited (M) = ... gm.

(*iii*) Mass of the copper deposited on the copper plate, $M - m = m'$ = ... gm.

(*iv*) Time for which copper deposited (t) = ... sec.

TABLE 1

S. No.	*Current in one direction*		*Current in opposite direction*	
	Time (min)	*Deflection*	*Time (min)*	*Deflection*
1	0	45°	0	45°
2	5		5	
3	10		10	
4	15		15	

Calculations

$$Z = \frac{m}{it}$$

where i is the current in the circuit and equal to $k\tan\theta$. In this experiment, we have kept $\theta = 45°$ constant and as such $i = k$, i.e. equal to the reduction factor of the Helmholtz galvanometer, so

$$Z = \frac{m}{kt} \text{ gm/coulomb}$$

$$= \ldots \text{ gm/coulomb}$$

Result

The electrochemical equivalent of copper (Z) is obtained

$= \ldots$ gm/coulomb

Precautions

(*i*) Deflection in the galvanometer be kept constant throughout the experiment.

(*ii*) Care must be taken during the measurement of the weight of copper plate.

Questions with Answers

1. What do you understand by electrolysis?

Ans. The decomposition of salt solution by passing a current through it is called electrolysis. During this process cation (positive ions) are liberated at cathode and anions (negative ions) are liberated at anode.

2. What is *ECE* of an element?

Ans. It is the mass of ions of an element liberated at an electrode when one ampere current is passed for one second through its electrolyte.

3. State the Faraday's laws of electrolysis?

Ans. (*i*) During electrolysis, the mass of ions liberated at any electrode is directly proportional to the charge Q flowing through the electrolyte. We have $m \propto Q$, or $m \propto it$, i.e. $m = Zit$ where Z is *ECE*.

(*ii*) When the same amount of charge is passed through several electrolytes, the amount of ions liberated at various electrodes are proportional to their chemical equivalent.

4. What is a voltameter?

Ans. This is a vessel in which the electrolysis is carried out.

5. What is a primary cell?

Ans. The chemical energy is converted into electrical energy in a primary cell. Due to some chemical reactions, e.m.f. is developed in this cell and these reactions are irreversible. Few examples of this type of cell are: Lechlanche cell, Daniel cell and Standard cadmium cell.

6. What is a secondary cell?

Ans. In these cells, electrical energy is first stored in the form of chemical energy and then chemical energy is converted into electrical energy..

7. What is basic difference between primary and secondary cells?

Ans. See Ans. 5, 6.

8. What is Helmholtz galvanometer?

Ans. This consists of two similar coils of same radius and of same number of turns. Each circular coil has a combination of three insulated copper wire coils having 2, 50 and 500 turns.

9. Why do we set the plane of the coil along the magnetic meridian?

Ans. The plane of the coil is set in the magnetic meridian so that the magnetic field produced by circular coil may be right angle to the horizontal component of earth's magnetic field.

10. Why do we adjust the deflection in the galvanometer nearly 45°?

Ans. Since $\tan\theta$ does not vary much near 45°. At this value of θ, a small error in the measurement does not affect our result.

Practical Tables

TABLE 1: WAVELENGTH AND REFRACTIVE INDEX FOR DIFFERENT COLOURS

Wavelength λ, (nm)	*Colour*	*Iceland spar*		*Quartz*	
		n_e	n_0	n_e	n_0
687	red	1.484	1.653	1.550	1.541
656	orange	1.485	1.655	1.551	1.542
589	yellow	1.486	1.658	1.553	1.544
527	green	1.489	1.664	1.556	1.547
486	blue	1.491	1.668	1.559	1.550
431	indigo	1.495	1.676	1.564	1.554
400	violet	1.498	1.683	1.568	1.558

TABLE 2: WAVELENGTH OF SPECTRAL LINES (Å)

Hydrogen	*Mercury*	*Neon*
6562.784 4861.327 4340.466 4101.736	4047 V 4078 V 4358 V 4916 bg	5765 y 5853 y 5882 o 6507 r
Sodium:	4960 g	*Cadmium:*
5890 D_1 5896 D_2	5461 g 5770 y	6438 5085
Helium:	5791 y	4799
3889 4026 4471 5876		4678 4662

TABLE 3: REFRACTIVE INDEX OF SUBSTANCES FOR λ = 5893 Å

Substance	*Ref. Index* (μ)	*Substance*	*Ref. Index* (μ)
Glass (Crown)	1.48–1.61	Alcohol	1.362
Glass (Flint)	1.53–1.96	Paraffin Oil	1.44
Glass (Soda)	1.50	Glycerine	1.473
Diamond	2.42	Kerosene	1.390
Ice	1.31	Castor Oil	1.472
Mica	1.56–1.61	Turpentine	1.47
Quartz	1.544	Olive Oil	1.46
Benzene	1.504	Linseed Oil	1.474
Water	1.33	Carbon disulphide	1.632
Canada balsam	1.53	Chloroform	1.449
Potash Alum	1.456	Plastic (Polysterine)	1.589
Coconut Oil	1.45	Liq. Paraffin	1.48
Rock Salt	1.544	Sod. Chloride	1.546
Ether	1.354		

TABLE 4: DISPERSIVE POWER (δ)

Type of Glass	*Dispersive Power* (ω)
Crown, Borosilicate	0.0155
Crown, Hard	0.0166
Crown, Soft	0.0177
Flint, Dense	0.0271
Flint, Light	0.0243
Flint, Extra dense	0.0330
Flint, Double extra dence	0.0360
Polysteren (Plastic) 15° C	0.0321

TABLE 5: OPTICAL ROTATORY POWER

Subseance	*Solvent*	*Specific Rotation (deg.)*
Cane Sugar, $C_{12}H_{22}O_{11}$	Water	+ 66.5
d-Camphor	Methyl alcohol	+ 43.7
d-Glucose	Water	+ 52.5
Invert Sugar	Water	– 19.7
Lactose	Water	+ 52.5
Nicotine	Water	– 79.4
Quinine	Ethyl alcohol	– 169.3
Rochelle Salt	Water	+ 29.7
Turpentine Oil		– 37.0

Specific rotation $\alpha = 100\ \theta/lc$, where θ is rotation in degree by liquid column of length l dm and conecntration C (in gm/100cc). All values at 20° C, for dextro-laevo rotation.

TABLE 6: RESISTANCES OF WIRE (R)

S.W.G. No.	*Diameer (mm)*	*Copper*	*Resustabce (Ω/m) Constantan*	*Manganin*
10	3.25	0.0021	0.057	0.051
12	2.64	0.0032	0.086	0.077
14	2.03	0.0054	0.146	0.131
16	1.63	0.0083	0.228	0.205
18	1.22	0.0148	0.495	0.361
20	0.914	0.0260	0.722	0.645
22	0.711	0.0235	1.20	1.07
24	0.559	0.070	1.93	1.73
26	0.457	0.105	0.89	2.58
28	0.374	0.155	4.27	3.82
30	0.315	0.222	6.08	5.45
32	0.274	0.293	8.02	7.18
34	0.234	0.404	11.1	9.9
36	0.193	0.590	16.2	14.5
38	0.152	0.950	26.2	23.2
40	0.122	1.48	40.6	36.3

TABLE 7: SPECIFIC RESISTANCE AND TEMPERATURE COEFFICIENT OF RESISTANCES

Substance	*Sp. Resistance* ($\times 10^{-6}$ *ohm*-cm)	*Temp. Coeff. of resitance* ($\times 10^{-4}$/°C)	*Substance*	*Sp. Resistance* ($\times 10^{-6}$ *ohm*-cm)	*Temp. Coeff. of resistance* ($\times 10^{-4}$/°C)
Aluminium	3.21	38.0	Manganin	44.5	–0.1 to 14.0
Copper	1.70	43.0	Nikel	11.8	27
Brass	6.3	10.0	Tungsten	5.5	51
Constantan	49	–0.4 to 0.1	Mercury	95.8	9.0
German Silver	33	2.5 to 6.0	Graphite	3000	–7.0
Iron	13	62	Carbon filament	4000	negative
Steel	19	16 to 22	Phosphor bronze	5 to 10	–
Silver	1.63	40	Quartz	1.2×10^{21}	–
Platinum	11	37	Mica	9×10^{21}	–
Gold	2.4	40	Zine	6.1	37
Nichrome	110	1.70	Tin	11.3	45

TABLE 8: ELECTRO–CHEMICAL EQUIVALENT

Element	*Atomic weight*	*Valency*	*E.C.E. gm./coulomb*
Aluminium	26.97	3	0.0000935
Oxygen	16.00	2	0.0000829
Silver	107.88	1	0.0011180
Copper	63.57	2	0.0003294
Zine	65.38	2	0.0003383
Gold	197.20	3	0.0006809
Nickle	58.69	2	0.0000304
Hydrogen	1.007	1	0.0000105

TABLE 9: THERMO E.M.F. (E)

Thermocouple	*Thermo* e.m.f.
Copper-constantan	41.8 Microvolt/° C
Copper-Iron	8.6 Microvolt/° C
Antimony-Bismuth	113 Microvolt/° C

TABLE 10: VARIATIONS OF THERMO–E.M.F. WITH TEMPERATURE

Temperature of cold junction	0°C									
Temp. of hot junction.	0	10	20	30	40	50	60	70	80	90	100
Thermo-emf (in mV) for Copper-constantan	0	0.39	0.79	1.19	1.61	2.03	2.47	2.91	3.36	3.81	4.28
Therom-emf (in mV) for Iron-constantan	0	0.52	1.05	1.58	2.12	2.66	3.20	3.75	4.30	4.85	5.40

TABLE 11: INTERNAL RESISTANCE AND E.M.F. OF CELLS

Cell	*Internal resistance (ohms)*	*E.M.F. (volts)*
Cadmium	900 Ω (very high)	1.0183 volt at 20°C
Daniel	3–4 Ω (fairly constant)	1.08
Leclanche	High, increases with usage	
Alkali	Low	1.46
Accumulator	Very Low	2.1
Lead acumulator		

TABLE 12: PERMITTIVITIES (RELATIVE VALUES)

Dielectric	ε	*Dielectric*	ε
Water	81	Mica	7.5
Air	1.00058	Alcohol	26
Kerosene	2.0	Glass	6.0
Pareffin	2.0	Porcelain	6.0
Plexiglas	3.5	Ebonite	2.7
Polyethylene	2.3		

TABLE 13: MAGNETIC ELEMENTS

Station	*Dip*	*H oersted*	*Satation*	*Dip*	*H oersted*
Agra	40°41'	0.348	Gorakhpur	39°40'	0.358
Aligarh	41°50'	0.346	Jaipur	40°30'	0.347
Allahabad	37°10'	0.353	Kanupr	38°39'	0.363
Delhi	42°52'	0.345	Khurja	42°10'	0.343
Meerut	43°30'	0.339	Lucknow	40°00'	0.354
Varanasi	37°10'	0.364	Bareilly	42°20'	0.344
Dehra Dun	45°50'	0.332	Bombay	25°30'	0.376
Gwalior	39°00'	0.353	Calcutta	31°30'	0.382

TABLE 14: MAGMETIC SUSEPTIBILITIES OF PARA- AND DIA-MAGNETICS

Paramagnetic substance	$\mu - 1, 10^{-6}$	*Diamagnetic substance*	$\mu - 1, 10^{-6}$
Nitrogen	0.013	Hydrogen	−0.063
Ari	0.38	Benzenc	−7.5
Oxygen	1.9	Water	−9.0
Ebonite	14	Copper	−10.3
Aluminium	23	Glass	−12.6
Tangsten	176	Rock-salt	−12.6
Platinum	360	Quartz	−15.1
Liquid oxygen	3400	Bismuth	−176

TABLE 15: BANDGAP IN SEMICONDUCTORS

Substance	ΔE_g (eV)	*Substance*	ΔE_g (eV)
Diamond	7.2	In Sb	0.18
Silicon	1.12	Te	0.38
Germanium	0.67	Pbs	0.37
Ga As	1.43	Pb Te	0.32
Ga Sb	0.7		

LOGARITHMS

	0	1	2	3	4	5	6	7	8	9	1	2	3	4	5	6	7	8	9
10	0000	0043	0086	0128	0170						5	9	13	17	21	26	30	34	38
						0212	0253	0294	0334	0374	4	8	12	16	20	24	28	32	36
11	0414	0453	0492	0531	0569						4	8	12	16	20	23	27	31	35
						0607	0645	0682	0719	0755	4	7	11	15	18	22	26	29	33
12	0792	0828	0864	0899	0934						3	7	11	14	18	21	25	28	32
						0969	1004	1038	1072	1106	3	7	10	14	17	20	24	27	31
13	1139	1173	1206	1239	1271						3	6	10	13	16	19	23	26	29
						1303	1335	1367	1399	1430	3	7	10	13	16	19	22	25	29
14	1461	1492	1523	1533	1584						3	6	9	12	15	19	22	25	28
						1614	1644	1673	1703	1732	3	6	9	12	14	17	20	23	26
15	1761	1790	1818	1847	1875						3	6	9	11	14	17	20	23	26
						1903	1931	1959	1987	2014	3	6	8	11	14	17	19	22	25
16	2041	2068	2095	2122	2148						3	6	8	11	14	16	19	22	24
						2175	2201	2227	2253	2279	3	5	8	10	13	16	18	21	23
17	2304	2330	2355	2380	2405						3	5	8	10	13	15	18	20	23
						2430	2455	2480	2504	2529	3	5	8	10	12	15	17	20	22
18	2553	2577	2601	2625	2648						2	5	7	9	12	14	17	19	21
						2672	2695	2718	2742	2765	2	4	7	9	11	14	16	18	21
19	2788	2810	2833	2856	2878						2	7	7	9	11	13	16	18	20
						2900	2923	2945	2967	2989	2	4	6	8	11	13	15	17	19
20	3010	3032	3054	3075	3096	3118	3139	3160	3181	3201	2	4	6	8	11	13	15	17	19
21	3222	3243	3263	3284	3304	3324	3345	3365	3385	3404	2	4	6	8	10	12	14	16	18
22	3424	3444	3464	3483	3502	3522	3541	3560	3579	3598	2	4	6	8	10	12	14	15	17
23	3617	3636	3655	3674	3692	3711	3729	3747	3766	3784	2	4	6	7	9	11	13	15	17
24	3802	3820	3838	3856	3874	3892	3909	3927	3945	3962	2	4	5	7	9	11	12	14	16
25	3979	3997	4014	4031	4048	4065	4082	4099	4116	4133	2	3	5	7	9	10	12	14	15
26	4150	4166	4183	4200	4216	4232	4249	4265	4281	4298	2	3	5	7	8	10	11	13	15
27	4314	4330	4346	4362	4378	4393	4409	4425	4440	4456	2	3	5	6	8	9	11	13	14
28	4472	4487	4502	4518	4533	4548	4564	4579	4594	4609	2	3	5	6	8	9	11	12	14
29	4624	4639	4654	4669	4683	4698	4713	4728	4742	4757	1	3	4	6	8	9	10	12	13
30	4771	4786	4800	4814	4829	4843	4857	4871	4886	4900	1	3	4	6	7	9	10	11	13
31	4941	4928	4942	4955	4969	4983	4997	5011	5024	5038	1	3	4	6	7	8	10	11	12
32	5051	5065	5079	5092	5105	5119	5132	5145	5159	5172	1	3	4	5	7	8	9	11	12
33	5185	5198	5211	5224	5237	5250	5263	5276	5289	5302	1	3	4	5	6	8	9	10	12
34	5315	5328	5340	5353	5366	5378	5391	5403	5416	5428	1	3	4	5	6	8	9	10	11
35	5441	5453	5465	5478	5490	5502	5514	5527	5539	5551	1	2	4	5	6	7	9	10	11
36	5563	5575	5587	5599	5611	5623	5635	5647	5658	5670	1	2	4	5	6	7	8	10	11
37	5682	5694	5705	5717	5729	5740	5752	5763	5775	5786	1	2	3	5	6	7	8	9	10
38	5798	5809	5821	5832	5843	5855	5866	5877	5888	5899	1	2	3	5	6	7	8	9	10
39	5911	5922	5933	5944	5955	5966	5977	5988	5999	6010	1	2	3	4	5	7	8	9	10
40	6021	6031	6042	6053	6064	6075	6085	6096	6107	6117	1	2	3	4	5	6	8	9	10
41	6128	6138	6149	6160	6170	6180	6191	6201	6212	6222	1	2	3	4	5	6	7	8	9
42	6232	6243	6253	6263	6274	6284	6294	6304	6314	6325	1	2	3	4	5	6	7	8	9
43	6335	6345	6355	6365	6375	6385	6395	6405	6415	6425	1	2	3	4	5	6	7	8	9
44	6435	6444	6454	6464	6474	6484	6463	6503	6513	6522	1	2	3	4	5	6	7	8	9
45	6532	6542	6551	6561	6571	6580	6590	6599	6609	6618	1	2	3	4	5	6	7	8	9
46	6628	6637	6646	6656	6665	6675	6684	6693	6702	6712	1	2	3	4	5	6	7	7	8
47	6721	6730	6730	6749	6758	6767	6776	6785	6794	6803	1	2	3	4	5	5	6	7	8
48	6812	6821	6830	6839	6848	6857	6866	6875	6884	6893	1	2	3	4	4	5	6	7	8
49	6902	6911	6920	6928	6937	6946	6955	6964	6972	6981	1	2	3	4	4	5	6	7	8

LOGARITHMS

	0	1	2	3	4	5	6	7	8	9	1	2	3	4	5	6	7	8	9
50	6990	6998	7007	7016	7024	7033	7042	7050	7059	7067	1	2	3	3	4	5	6	7	8
51	7076	7084	7093	7101	7110	7118	7126	7135	7143	7152	1	2	3	3	4	5	6	7	8
52	7160	7168	7177	7185	7193	7202	7210	7218	7226	7235	1	2	2	3	4	5	6	7	7
53	7243	7251	7259	7267	7275	7284	7292	7300	7308	7316	1	2	2	3	4	5	6	6	7
54	7324	7332	7340	7348	7356	7364	7372	7380	7388	7396	1	2	2	3	4	5	6	6	7
55	7404	7412	/419	7427	7435	7443	7451	7459	7466	7474	1	2	2	3	4	5	5	6	7
56	7482	7490	7497	7505	7513	7520	7528	7536	7543	7551	1	2	2	3	4	5	5	6	7
57	7559	7566	7574	7582	7589	7597	7604	7612	7619	7627	1	2	2	3	4	5	5	6	7
58	7634	7642	7649	7657	7664	7672	7679	7686	7694	7701	1	1	2	3	4	4	5	6	7
59	7709	7716	7723	7731	.7738	7745	7752	7760	7767	7774	1	1	2	3	4	4	5	6	7
60	7782	7789	7796	7803	7810	7818	7825	7832	7839	7846	1	1	2	3	4	4	5	6	6
61	7853	7860	7868	7875	7882	7889	7896	7903	7910	7917	1	1	2	3	4	4	5	6	6
62	7924	7931	7938	7945	7952	7959	7966	7973	7980	7987	1	1	2	3	3	4	5	6	6
63	7993	8000	8007	8014	8021	8028	8035	8041	8048	8055	1	1	2	3	3	4	5	5	6
64	8062	8069	8075	8082	8089	8096	8102	8109	8116	8122	1	1	2	3	3	4	5	5	6
65	8129	8136	8142	8149	8156	8162	8169	8176	8182	8189	1	1	2	3	3	4	5	5	6
66	8195	8202	8209	8215	8222	8228	8235	8241	8248	8254	1	1	2	3	3	4	5	5	6
67	8261	8267	8274	8280	8287	8293	8299	8306	8312	8319	1	1	2	3	3	4	5	5	6
68	8325	8331	8338	8344	8351	8357	8363	8370	8376	8382	1	1	2	3	3	4	4	5	6
69	8388	8395	8401	8407	8414	8420	8426	8432	8439	8445	1	1	2	2	3	4	4	5	6
70	8451	8457	8463	8470	8476	8482	8488	8494	8500	8506	1	1	2	2	3	4	4	5	6
71	8513	8519	8525	8531	8537	8543	8549	8555	8561	8567	1	1	2	2	3	4	4	5	5
72	8573	8579	8585	8591	8597	8603	8609	8615	8621	8627	1	1	2	2	3	4	4	5	5
73	8633	8639	8645	8651	8657	8663	8669	8675	8681	8686	1	1	2	2	3	4	4	5	5
74	8692	8698	8704	8710	8716	8722	8727	8733	8739	8745	1	1	2	2	3	4	4	5	5
75	8751	8756	8762	8768	8774	8779	8785	8791	8797	8802	1	1	2	2	3	3	4	5	5
76	8808	8814	8820	8825	8831	8837	8842	8848	8854	8859	1	1	2	2	3	3	4	5	5
77	8865	8871	8876	8882	8887	8893	8899	8904	8910	8915	1	1	2	2	3	3	4	4	5
78	8921	8972	8932	8938	8943	8949	8954	8960	8965	8971	1	1	2	2	3	3	4	4	5
79	8976	8982	8987	8993	8998	9004	9009	9015	9020	9025	1	1	2	2	3	3	4	4	5
80	9031	9036	9042	9047	9053	9058	9063	9069	9074	9079	1	1	2	2	3	3	4	4	5
81	9085	9090	9096	9101	9106	9112	9117	9122	9128	9133	1	1	2	2	3	3	4	4	5
82	9138	9143	9149	9154	9159	9165	9170	9175	9180	9186	1	1	2	2	3	3	4	4	5
83	9191	9196	9201	9206	9212	9217	9222	9227	9232	9238	1	1	2	2	3	3	4	4	5
84	9243	9248	9253	9258	9263	9269	9274	9279	9284	9289	1	1	2	2	3	3	4	4	5
85	9294	9299	9204	9309	9315	9320	9325	9330	9335	9340	1	1	2	2	3	3	4	4	5
86	9345	9350	9355	9360	9365	9370	9375	9380	9385	9390	1	1	2	2	3	3	4	4	5
87	9395	9400	9405	9410	9415	9420	9425	9430	9435	9440	0	1	1	2	2	3	3	4	4
88	9445	9450	9455	9460	9465	9469	9474	9479	9484	9489	0	1	1	2	2	3	3	4	4
89	9494	9499	9504	9509	9513	9518	9523	9528	9533	9538	0	1	1	2	2	3	3	4	4
90	9542	9547	9452	9557	9562	9566	9571	9576	9581	9586	0	1	1	2	2	3	3	4	4
91	9590	9595	9600	9605	9609	9614	9619	9624	9628	9633	0	1	1	2	2	3	3	4	4
92	9638	9643	9647	9652	9657	9661	9666	9671	9675	9680	0	1	1	2	2	3	3	4	4
93	9685	9689	9694	9699	9703	9708	9713	9717	9722	9727	0	1	1	2	2	3	3	4	4
94	9731	9736	9741	9745	9750	9754	9759	9763	9768	9773	0	1	1	2	2	3	3	4	4
95	9777	9782	9786	9791	9795	9800	9805	9809	9814	9818	0	1	1	2	2	3	3	4	4
96	9823	9827	9832	9836	9841	9845	9845	9854	9859	9863	0	1	1	2	2	3	3	4	4
97	9868	9872	9877	9881	9886	9890	9894	9899	9903	9908	0	1	1	2	2	3	3	4	4
98	9912	9917	9921	9926	9930	9934	9939	9943	9948	9952	0	1	1	2	2	3	3	4	4
99	9956	9961	9965	9969	9974	9978	9983	9987	9991	9996	0	1	1	2	2	3	3	3	4

ANTILOGARITHMS

	0	1	2	3	4	5	6	7	8	9	1	2	3	4	5	6	7	8	9
.00	1000	1002	1005	1007	1009	1012	1014	1016	1019	1021	0	0	1	1	1	1	2	2	2
.01	1023	1026	1028	1030	1033	1035	1038	1040	1042	1045	0	0	1	1	1	1	2	2	2
.02	1047	1050	1052	1054	1057	1059	1062	1064	1067	1069	0	0	1	1	1	1	2	2	2
.03	1072	1074	1076	1079	1081	1084	1086	1089	1091	1094	0	0	1	1	1	1	2	2	2
.04	1096	1099	1102	1104	1107	1109	1112	1114	1117	1119	0	1	1	1	1	2	2	2	2
.05	1122	1125	1127	1130	1132	1135	1138	1140	1143	1146	0	1	1	1	1	2	2	2	2
.06	1148	1151	1153	1156	1159	1161	1164	1167	1169	1172	0	1	1	1	1	2	2	2	2
.07	1175	1178	1180	1183	1186	1189	1191	1194	1197	1199	0	1	1	1	1	2	2	2	2
.08	1202	1205	1208	1211	1213	1216	1219	1222	1225	1227	0	1	1	1	1	2	2	2	3
.09	1230	1233	1236	1239	1242	1245	1247	1250	1253	1256	0	1	1	1	1	2	2	2	3
.10	1259	1262	1265	1268	1271	1274	1276	1279	1282	1285	0	1	1	1	1	2	2	2	3
.11	1288	1291	1294	1297	1300	1303	1306	1309	1312	1315	0	1	1	1	2	2	2	2	3
.12	1318	1321	1324	1327	1330	1334	1337	1340	1343	1346	0	1	1	1	2	2	2	2	3
.13	1349	1352	1355	1358	1361	1365	1368	1371	1374	1377	0	1	1	1	2	2	2	3	3
.14	1380	1384	1387	1390	1393	1396	1400	1403	1406	1409	0	1	1	1	2	2	2	3	3
.15	1413	1416	1419	1422	1426	1429	1432	1435	1439	1442	0	1	1	1	2	2	2	3	3
.16	1445	1449	1452	1455	1459	1462	1466	1469	1472	1476	0	1	1	1	2	2	2	3	3
.17	1479	1483	1486	1489	1493	1496	1500	1503	1507	1510	0	1	1	1	2	2	2	3	3
.18	1514	1517	1521	1524	1528	1531	1535	1538	1542	1545	0	1	1	1	2	2	2	3	3
.19	1549	1552	1556	1560	1563	1567	1570	1574	1578	1581	0	1	1	1	2	2	3	3	3
.20	1585	1589	1592	1596	1600	1603	1607	1611	1614	1618	0	1	1	1	2	2	3	3	3
.21	1622	1626	1629	1633	1637	1641	1644	1648	1652	1656	0	1	1	2	2	2	3	3	3
.22	1660	1663	1667	1671	1675	1679	1683	1687	1690	1694	0	1	1	2	2	2	3	3	3
.23	1698	1702	1706	1710	1714	1718	1722	1726	1730	1734	0	1	1	2	2	2	3	3	4
.24	1738	1742	1746	1750	1754	1758	1762	1766	1770	1774	0	1	1	2	2	2	3	3	4
.25	1778	1782	1786	1791	1795	1799	1803	1807	1811	1816	0	1	1	2	2	2	3	3	4
.26	1820	1824	1828	1832	1837	1841	1845	1849	1854	1858	0	1	1	2	2	3	3	3	4
.27	1862	1866	1871	1875	1879	1884	1888	1892	1897	1901	0	1	1	2	2	3	3	3	4
.28	1905	1910	1914	1919	1923	1928	1932	1936	1941	1945	0	1	1	2	2	3	3	4	4
.29	1950	1954	1959	1963	1968	1972	1977	1982	1986	1991	0	1	1	2	2	3	3	4	4
.30	1995	2000	2004	2009	2014	2018	2023	2028	2032	2037	0	1	1	2	2	3	3	4	4
.31	2042	2046	2051	2056	2061	2065	2070	2075	2080	2084	0	1	1	2	2	3	3	4	4
.32	2089	2094	2099	2104	2109	2113	2118	2123	2128	2133	0	1	1	2	2	3	3	4	4
.33	2138	2143	2148	2153	2158	2163	2168	2173	2178	2183	0	1	1	2	2	3	3	4	4
.34	2188	2193	2198	2203	2208	2213	2218	2223	2228	2234	1	1	2	2	3	3	4	4	5
.35	2239	2244	2249	2254	2259	2265	2270	2275	2280	2286	1	1	2	2	3	3	4	4	5
.36	2291	2296	2301	2307	2312	2317	2323	2328	2333	2339	1	1	2	2	3	3	4	4	5
.37	2344	2350	2355	2360	2366	2371	2377	2382	2388	2393	1	1	2	2	3	3	4	4	5
.38	2399	2404	24010	2415	2421	2427	2432	2438	2473	2449	1	1	2	2	3	3	4	4	5
.39	2455	2460	2466	2472	2477	2483	2489	2495	2500	2506	1	1	2	2	3	3	4	5	5
.40	2512	2518	2523	2529	2535	2541	2547	2553	2559	2564	1	1	2	2	3	4	4	5	5
.41	2570	2576	2582	2583	2594	2600	2606	2612	2618	2624	1	1	2	2	3	4	4	5	5
.42	2630	2636	2642	2649	2655	2661	2667	2673	2679	2685	1	1	2	2	3	4	4	5	6
.43	2692	2698	2704	2710	2716	2723	2729	2735	2742	2748	1	1	2	2	3	4	4	5	6
.44	2754	2761	2767	2773	2780	2786	2793	2799	2805	2812	1	1	2	3	3	4	4	5	6
.45	2818	2825	2831	2838	2844	2815	2858	2864	2871	2877	1	1	2	3	3	4	5	5	6
.46	2884	2891	2897	2904	2911	2917	2924	2931	2938	2944	1	1	2	3	3	4	5	5	6
.47	2951	2958	2965	2972	2979	2985	2992	2999	3006	3013	1	1	2	3	3	4	5	5	6
.48	3020	3027	3034	3041	3048	3055	3062	3069	3076	3083	1	1	2	3	4	4	5	6	6
.49	3090	3097	3105	3112	3119	3126	3133	3141	3148	3155	1	1	2	3	4	4	5	6	6

ANTILOGARITHMS

	0	1	2	3	4	5	6	7	8	9	1	2	3	4	5	6	7	8	9
.50	3162	3170	3177	3184	3192	3199	3206	3214	3221	3228	1	1	2	3	4	4	5	6	7
.51	3236	3243	3251	3258	3266	3273	3281	3289	3296	3304	1	2	2	3	4	5	5	6	7
.52	3311	3319	3327	3334	3342	3350	3357	3365	3373	3381	1	2	2	3	4	5	5	6	7
.53	3388	3396	3404	3412	3420	3428	3436	3443	3451	3459	1	2	2	3	4	5	6	6	7
.54	3467	3475	3483	3491	3499	3508	3516	3524	3532	3540	1	2	2	3	4	5	6	6	7
.55	3548	3556	3565	3573	3581	3589	3597	3606	3614	3622	1	2	2	3	4	5	6	7	7
.56	3631	3639	3648	3656	3664	3673	3681	3690	3698	3707	1	2	3	3	4	5	6	7	8
.57	3715	3724	3733	3741	3750	3758	3767	3776	3784	3793	1	2	3	3	4	5	6	7	8
.58	3802	3811	3819	3828	3837	3846	3855	3864	3873	3882	1	2	3	4	4	5	6	7	8
.59	3890	3899	3908	3917	3926	3936	3945	3954	3963	3972	1	2	3	4	5	5	6	7	8
.60	3981	3990	3999	4009	4018	4027	4036	4046	4055	4064	1	2	3	4	5	6	6	7	8
.61	4074	4083	4093	4102	4111	4121	4130	4140	4150	4159	1	2	3	4	5	6	7	8	9
.62	4169	4178	4188	4198	4207	4217	4227	4236	4246	4256	1	2	3	4	5	6	7	8	9
.63	4266	4276	4285	4295	4305	4315	4325	4335	4345	4355	1	2	3	4	5	6	7	8	9
.64	4365	4375	4385	4395	4406	4416	4426	4436	4446	4457	1	2	3	4	5	6	7	8	9
.65	4467	4477	4487	4498	4508	4519	4529	4539	4550	4560	1	2	3	4	5	6	7	8	9
.66	4571	4581	4592	4603	4613	4624	4634	4645	4656	4667	1	2	3	4	5	6	7	9	10
.67	4677	4688	4699	4710	4721	4732	4742	4753	4764	4775	1	2	3	4	5	7	8	9	10
.68	4786	4797	4808	4819	4831	4842	4853	4864	4875	4887	1	2	3	4	6	7	8	9	10
.69	4898	4909	4920	4932	4943	4955	4966	4977	4989	5000	1	2	3	5	6	7	8	9	10
.70	5012	5023	5035	5047	5058	5070	5082	5093	5105	5117	1	2	4	5	6	7	8	9	11
.71	5129	5140	5152	5164	5176	5188	5200	5212	5224	5236	1	2	4	5	6	7	8	10	11
.72	5248	5260	5272	5284	5297	5309	5321	5333	5346	5358	1	2	4	5	6	7	9	10	11
.73	5370	5383	5395	5408	5420	5433	5445	5458	5470	5483	1	3	4	5	6	8	9	10	11
.74	5495	5508	5521	5534	5546	5559	5572	5585	5598	5610	1	3	4	5	6	8	9	10	12
.75	5623	5636	5649	5662	5675	5689	5702	5715	5728	5741	1	3	4	5	7	8	9	10	12
.76	5754	5768	5781	5794	5808	5821	5834	5848	5861	5875	1	3	4	5	7	8	9	11	12
.77	5888	5902	5916	5929	5943	5957	5970	5984	5998	6012	1	3	4	5	7	8	10	11	12
.78	6026	6039	6053	6067	6081	6095	6109	6124	6138	6152	1	3	4	6	7	8	10	11	13
.79	6166	6180	6194	6209	6223	9237	6252	6266	6281	6295	1	3	4	6	7	9	10	11	13
.80	6310	6324	6339	6353	6368	6383	6397	6412	6427	6442	1	3	4	6	7	9	11	12	13
.81	6457	6471	6486	6501	6516	6531	6546	6561	6577	6592	2	3	5	6	8	9	11	12	14
.82	6607	6622	6637	6653	6668	6683	6699	6714	6730	6745	2	3	5	6	8	9	11	12	14
.83	6761	6776	6792	6808	6823	6839	6855	6871	6887	6902	2	3	5	6	8	9	10	13	14
.84	6918	6934	6950	6966	6982	6998	7015	7031	7047	7063	2	3	5	6	8	10	11	13	15
.85	7079	7096	7112	7129	7145	7161	7178	7194	7211	7228	2	3	5	7	8	10	12	13	15
.86	7244	7261	7278	7295	7311	7328	7345	7362	7379	7396	2	3	5	7	8	10	12	13	15
.87	7413	7430	7447	7464	7482	7499	7516	7534	7551	7568	2	3	5	7	9	10	12	14	16
.88	7586	7603	7621	7638	7656	7674	7691	7709	7727	7745	2	4	5	7	9	11	12	14	16
.89	7762	7780	7798	7816	7834	7852	7870	7889	7907	7925	2	4	5	7	9	11	13	14	16
.90	7943	7962	7980	7998	8017	8035	8054	8072	8091	8110	2	4	6	7	9	11	13	15	17
.91	8128	8147	8166	8185	8204	8222	8241	8260	8279	8299	2	4	6	8	9	11	13	15	17
.92	8318	8337	8356	8375	8395	8414	8433	8453	8472	8492	2	4	6	8	10	12	14	15	17
.93	8511	8531	8551	8570	8590	8610	8630	8650	8670	8690	2	4	6	8	10	12	14	16	18
.94	8710	8730	8750	8770	8790	8810	8831	8851	8872	8892	2	4	6	8	10	12	14	16	18
.95	8913	8933	8954	8974	7995	9016	9036	9057	9078	9099	2	4	6	8	10	12	15	17	19
.96	9120	9141	9162	9183	9204	9226	9247	9268	9290	9311	2	4	6	8	11	13	15	17	19
.97	9333	9354	9376	9397	9419	9441	9462	9484	9506	9528	2	4	7	9	11	13	15	17	20
.98	9550	9572	9594	9616	9638	9661	9683	9705	9727	9750	2	4	7	9	11	13	16	18	20
.99	9772	9795	9817	9840	9863	9886	9908	9931	9954	9977	2	5	7	9	11	14	16	17	20

SINES AND COSINES

Deg-rees	0°	6°	12°	18°	24°	30°	36°	42°	48°	54°	60°	
0	0.0000	0.0017	0.0035	0.0052	0.0070	0.0087	0.0105	0.0122	0.0140	0.0157	0.0175	89
1	0.0175	0.0192	0.0209	0.0227	0.0244	0.0262	0.0279	0.0297	0.0314	0.0332	0.0349	88
2	0.0349	0.0366	0.0384	0.0401	0.0419	0.0436	0.0454	0.0471	0.0488	0.0506	0.0523	87
3	0.0523	0.0541	0.0558	0.0576	0.0593	0.0610	0.0628	0.0645	0.0663	0.0680	0.0698	86
4	0.0698	0.0715	0.0732	0.0750	0.0767	0.0785	0.0802	0.0819	0.0837	0.0854	0.0872	85
5	0.0872	0.0889	0.0906	0.0924	0.0941	0.0958	0.0970	0.0993	0.1011	0.1028	0.1045	84
6	0.1045	0.1063	0.1080	0.1097	0.1115	0.1132	0.1149	0.1167	0.1184	0.1201	0.1219	83
7	0.1219	0.1236	0.1253	0.1271	0.1288	0.1305	0.1323	0.1340	0.1357	0.1374	0.1392	82
8	0.1392	0.1409	0.1426	0.1444	0.1461	0.1478	0.1495	0.1513	0.1530	0.1547	0.1564	81
9	0.1564	0.1582	0.1599	0.1616	0.1633	0.1650	0.1668	0.1685	0.1702	0.1719	0.1736	80
10	0.1736	0.1754	0.1771	0.1788	0.1805	0.1822	0.1840	0.1857	0.1874	0.1891	0.1908	79
11	0.1908	0.1925	0.1942	0.1959	0.1977	0.1994	0.2011	0.2028	0.2045	0.2062	0.2079	78
12	0.2079	0.2096	0.2113	0.2130	0.2147	0.2164	0.2181	0.2198	0.2215	0.2233	0.2250	77
13	0.2250	0.2267	0.2284	0.2300	0.2317	0.2334	0.2351	0.2368	0.2385	0.2402	0.2419	76
14	0.2419	0.2436	0.2453	0.2470	0.2487	0.2504	0.2521	0.2538	0.2554	0.2571	0.2588	75
15	0.2588	0.2605	0.2622	0.2639	0.2656	0.2672	0.2689	0.2706	0.2723	0.2740	0.2756	74
16	0.2756	0.2773	0.2790	0.2807	0.2823	0.2840	0.2857	0.2874	0.2890	0.2907	0.2924	73
17	0.2924	0.2940	0.2957	0.2974	0.2990	0.3007	0.3024	0.3040	0.3057	0.3074	0.3090	72
18	0.3090	0.3107	0.3123	0.3140	0.3156	0.3173	0.3190	0.3206	0.3223	0.3239	0.3256	71
19	0.3256	0.3272	0.3289	0.3305	0.3322	0.3338	0.3355	0.3371	0.3387	0.3404	0.3420	70
20	0.3420	0.3437	0.3453	0.3469	0.3486	0.3502	0.3518	0.3535	0.3551	0.3567	0.3584	69
21	0.3584	0.3600	0.3616	0.3633	0.3649	0.3665	0.3681	0.3697	0.3714	0.3730	0.3746	68
22	0.3746	0.3762	0.3778	0.3795	0.3811	0.3827	0.3843	0.3859	0.3875	0.3891	0.3907	67
23	0.3907	0.3923	0.3939	0.3955	0.3911	0.3987	0.4003	0.4019	0.4035	0.4051	0.4067	66
24	0.4067	0.4083	0.4099	0.4115	0.4131	0.4147	0.4163	0.4179	0.4195	0.4210	0.4226	65
25	0.4226	0.4242	0.4258	0.4274	0.4289	0.4305	0.4321	0.4337	0.4352	0.4368	0.4384	64
26	0.4384	0.4399	0.4415	0.4431	0.4446	0.4462	0.4478	0.4493	0.4509	0.4524	0.4540	63
27	0.4540	0.4555	0.4571	0.4586	0.4602	0.4617	0.4633	0.4648	0.4664	0.4679	0.4695	62
28	0.4695	0.4710	0.4726	0.4741	0.4756	0.4772	0.4787	0.4802	0.4818	0.4833	0.4848	61
29	0.4848	0.4863	0.4879	0.4894	0.4909	0.4924	0.4939	0.4955	0.4970	0.4985	0.5000	60
30	0.5000	0.5015	0.5030	0.5045	0.5060	0.5075	0.5090	0.5105	0.5120	0.5135	0.5150	59
31	0.5150	0.5165	0.5180	0.5159	0.5210	0.5225	0.5240	0.5255	0.5270	0.5284	0.5299	58
32	0.5299	0.5314	0.5329	0.5344	0.5358	0.5373	0.5388	0.5402	0.5417	0.5432	0.5446	57
33	0.5446	0.5461	0.5476	0.5490	0.5505	0.5519	0.5534	0.5548	0.5563	0.5577	0.5592	56
34	0.5592	0.5606	0.5621	0.5635	0.5650	0.5664	0.5678	0.5693	0.5707	0.5721	0.5736	55
35	0.5736	0.5750	0.5764	0.5779	0.5793	0.5807	0.5821	0.5835	0.5850	0.5864	0.5878	54
36	0.5878	0.5892	0.6906	0.5920	0.5934	0.5948	0.5962	0.5976	0.5990	0.6004	0.6018	53
37	0.6018	0.6032	0.6046	0.6060	0.6074	0.6088	0.6101	0.6115	0.6129	0.6143	0.6157	52
38	0.6157	0.6170	0.6184	0.6198	0.6211	0.6225	0.6239	0.6252	0.6266	0.6280	0.6293	51
39	0.6293	0.6307	0.6320	0.6334	0.6347	0.6361	0.6374	0.6388	0.6401	0.6414	0.6428	50
40	0.6428	0.6441	0.6455	0.6468	0.6481	0.6494	0.6508	0.6521	0.6534	0.6547	0.6561	49
41	0.6561	0.6574	0.6587	0.6600	0.6613	0.6626	0.6639	0.6652	0.6665	0.6678	0.6691	48
42	0.6691	0.6704	0.6717	0.6730	0.6743	0.6756	0.6769	0.6782	0.6794	0.6807	0.6820	47
43	0.6820	0.6833	0.6845	0.6858	0.6871	0.6884	0.6896	0.6909	0.6921	0.6934	0.6947	46
44	0.6947	0.6959	0.6972	0.6984	0.6997	0.7009	0.7022	0.7034	0.7046	0.7059	0.7071	45
	60°	54°	48°	42°	36°	30°	24°	18°	12°	6°	0°	Degrees

SINES AND COSINES

Deg-rees	0°	6°	12°	18°	24°	30°	36°	42°	48°	54°	60°	
45	0.7071	0.7083	0.7096	0.7108	0.7120	0.7233	0.7145	0.7157	0.7169	0.7181	0.7193	44fs
46	0.7193	0.7206	0.7218	0.7230	0.7242	0.7254	0.7266	0.7278	0.7290	0.7302	0.7314	43
47	0.7314	0.7325	0.7337	0.7349	0.7361	0.7373	0.7385	0.7396	0.7408	0.7420	0.7431	42
48	0.7431	0.7443	0.7455	0.7466	0.7478	0.7490	0.7501	0.7513	0.7524	0.7536	0.7547	41
49	0.7547	0.7559	0.7570	0.7581	0.9593	0.7604	0.7615	0.7627	0.7638	0.7649	0.7660	40
50	0.7660	0.7672	0.7683	0.7694	0.7705	0.7716	0.7727	0.7738	0.7749	0.7760	0.7771	39
51	0.7771	0.7782	0.7793	0.7804	0.7815	0.7826	0.7837	0.7848	0.7859	0.7869	0.7880	38
52	0.7880	0.7891	0.7902	0.7912	0.7923	0.7934	0.7944	0.7955	0.7965	0.7976	0.7986	37
53	0.7986	0.7997	0.8007	0.8018	0.8028	0.8039	0.8049	0.8059	0.8070	0.8080	0.8090	36
54	0.8090	0.8100	0.8111	0.8121	0.8131	0.8141	0.8151	0.8161	0.8171	0.8181	0.8192	35
55	0.8192	0.8202	0.8211	0.8221	0.8231	0.8241	0.8251	0.8261	0.8271	0.8281	0.8290	34
56	0.8290	0.8300	0.8310	0.8320	0.8329	0.8339	0.8341	0.8358	0.8368	0.8377	0.8387	33
57	0.8387	0.8396	0.8406	0.8415	0.8425	0.8434	0.8443	0.8453	0.8462	0.8471	0.8480	32
58	0.8480	0.8490	0.8499	0.8508	0.8517	0.8526	0.8536	0.8545	0.8554	0.8563	0.8572	31
59	0.8572	0.8581	0.8590	0.8599	0.8607	0.8616	0.8625	0.8634	0.8643	0.8652	0.8660	30
60	0.8660	0.8669	0.8678	0.8686	0.8695	0.8704	0.8712	0.8721	0.8729	0.8738	0.8746	29
61	0.8746	0.8755	0.8763	0.8771	0.8780	0.8788	0.8796	0.8805	0.8813	0.8821	0.8829	28
62	0.8829	0.8838	0.8846	0.8854	0.8862	0.8870	0.8878	0.8886	0.8894	0.8902	0.8910	27
63	0.8910	0.8918	0.8926	0.8934	0.8942	0.8949	0.8957	0.8965	0.8973	0.8980	0.8988	26
64	0.8988	0.8996	0.9003	0.9011	0.9018	0.9026	0.9033	0.9041	0.9048	0.9056	0.9063	25
65	0.9063	0.9070	0.9078	0.9085	0.9092	0.9100	0.9107	0.9114	0.2191	0.9128	0.9135	24
66	0.9135	0.9143	0.9150	0.9157	0.9164	0.9171	0.9178	0.9184	0.9191	0.9198	0.9205	23
67	0.9205	0.9212	0.9219	0.9225	0.9232	0.9239	0.9245	0.9252	0.9259	0.9265	0.9272	22
68	0.9272	0.9278	0.9285	0.9291	0.9298	0.9304	0.9311	0.9317	0.9323	0.9330	0.9336	21
69	0.9336	0.9342	0.9348	0.9354	0.9361	0.9367	0.9373	0.9379	0.9385	0.9391	0.9391	20
70	0.9397	0.9403	0.9409	0.9415	0.9421	0.9426	0.9433	0.9438	0.9444	0.9449	0.9455	19
71	0.9455	0.9461	0.9466	0.9472	0.9478	0.9483	0.9489	0.9494	0.9500	0.9505	0.9511	18
72	0.9511	0.9516	0.9521	0.9527	0.9532	0.9537	0.9542	0.9548	0.9553	0.9558	0.9563	17
73	0.9563	0.9568	0.9573	0.9578	0.9583	0.9588	0.9593	0.9598	0.9603	0.9608	0.9613	16
74	0.9613	0.9617	0.9622	0.9627	0.9632	0.9636	0.9641	0.9646	0.9650	0.9655	0.9659	15
75	0.9699	0.9664	0.9668	0.9673	0.9677	0.9681	0.9686	0.9690	0.9694	0.9699	0.9703	14
76	0.9703	0.9707	0.9711	0.9715	0.9720	0.9724	0.9728	0.9732	0.9736	0.9740	0.9744	13
77	0.9744	0.9748	0.9751	0.9755	0.9759	0.9763	0.9767	0.9770	0.9774	0.9778	0.9781	12
78	0.9781	0.9785	0.9789	0.9792	0.9796	0.9799	0.9803	0.9806	0.9810	0.9813	0.9816	11
79	0.9816	0.9820	0.9823	0.9826	0.9829	0.9833	0.9836	0.9839	0.9842	0.9845	0.9848	10
80	0.9848	0.9851	0.9854	0.9857	0.9860	0.9863	0.9866	0.9869	0.9871	0.9874	0.9877	9
81	0.9877	0.9880	0.9882	0.9885	0.9888	0.9190	0.9893	0.9895	0.9898	0.9900	0.9903	8
82	0.9903	0.9905	0.9907	0.9910	0.9912	0.9914	0.9917	0.9919	0.9921	0.9923	0.9925	7
83	0.9925	0.9928	0.9930	0.9932	0.9934	0.9936	0.9918	0.9940	0.9942	0.9943	0.9945	6
84	0.9945	0.9947	0.9949	0.9951	0.9952	0.9954	0.9956	0.9957	0.9959	0.9960	0.9962	5
85	0.9962	0.9963	0.9965	0.9966	0.9968	0.9969	0.9971	0.9972	0.9973	0.9974	0.9976	4
86	0.9976	0.9977	0.9978	0.9979	0.9980	0.9981	0.9982	0.9983	0.9984	0.9985	0.9986	3
87	0.9986	0.9987	0.9988	0.9989	0.9990	0.9990	0.9991	0.9992	0.9993	0.9993	0.9994	2
88	0.9994	0.9995	0.9995	0.9996	0.9996	0.9997	0.9997	0.9997	0.9998	0.9998	0.9998	1
89	0.9998	0.9999	0.9999	0.9999	0.9999	1.0000	1.0000	1.0000	1.0000	1.0000	1.0000	0
	60°	54°	48°	42°	36°	30°	24°	18°	12°	6°	0°	Degress

TANGENTS AND COTANGENTS

Degrees	0°	10°	20°	30°	40°	50°	60°	1°	2°	3°	4°	5°	
0	0.0000	0.0029	0.0058	0.0087	0.0116	0.0145	0.0175	3	6	9	12	14	89
1	0.0175	0.0204	0.0233	0.0262	0.0291	0.0320	0.0349	3	6	9	12	15	88
2	0.0349	0.0378	0.0407	0.0437	0.0466	0.0495	0.0524	3	6	9	12	15	87
3	0.0524	0.0553	0.0582	0.0612	0.0641	0.0670	0.0699	3	6	9	12	15	86
4	0.0699	0.0729	0.0758	0.0787	0.0816	0.0846	0.0875	3	6	9	12	15	85
5	0.0875	0.0904	0.0934	0.0963	0.0992	0.1022	0.1051	3	6	9	12	15	84
6	0.1051	0.1080	0.1110	0.1139	0.1169	0.1198	0.1228	3	6	9	12	15	83
7	0.1228	0.1257	0.1287	0.1317	0.1346	0.1376	0.1405	3	6	9	12	15	82
8	0.1405	0.1435	0.1465	0.1495	0.1524	0.1554	0.1584	3	6	9	12	15	81
9	0.1584	0.1614	0.1644	0.1673	0.1703	0.1733	0.1763	3	6	9	12	15	80
10	0.1763	0.1793	0.1823	0.1853	0.1883	0.1914	0.1944	3	6	9	12	15	79
11	0.1914	0.1974	0.2004	0.2035	0.2065	0.2095	0.2126	3	6	9	12	15	78
12	0.2126	0.2156	0.2186	0.2217	0.2247	0.2278	0.2309	3	6	9	12	15	77
13	0.2309	0.2339	0.2370	0.2401	0.2432	0.2462	0.2493	3	6	9	12	15	76
14	0.2493	0.2524	0.2555	0.2586	0.2617	0.2648	0.2679	3	6	9	12	16	75
15	0.2679	0.2711	0.2742	0.2773	0.2805	0.2836	0.2867	3	6	9	13	16	74
16	0.2867	0.2899	0.2931	0.2962	0.2994	0.3026	0.3057	3	6	9	13	16	73
17	0.3057	0.3089	0.3121	0.3153	0.3185	0.3217	0.3249	3	6	10	13	16	72
18	0.3249	0.3281	0.3314	0.3346	0.3378	0.3411	0.3443	3	6	10	13	16	71
19	0.3443	0.3476	0.3508	0.3541	0.3574	0.3607	0.3640	3	6	10	13	17	70
20	0.3640	0.3673	0.3706	0.3739	0.3772	0.3805	0.3839	3	7	10	13	17	69
21	0.3839	0.3872	0.3906	0.3939	0.3973	0.4006	0.4040	3	7	10	13	17	68
22	0.4040	0.4074	0.4108	0.4142	0.4176	0.4210	0.4245	3	7	10	14	17	67
23	0.4245	0.4279	0.4314	0.4348	0.4383	0.4417	0.4452	3	7	10	14	17	66
24	0.4452	0.4487	0.4522	0.4557	0.4592	0.4628	0.4663	3	7	10	14	18	65
25	0.4663	0.4899	0.4734	0.4770	0.4806	0.4841	0.4877	4	7	11	14	18	64
26	0.4877	0.4913	0.4950	0.4986	0.5022	0.5059	0.5095	4	7	11	15	18	63
27	0.5095	0.5132	0.5169	0.5206	0.5243	0.5280	0.5317	4	8	11	15	18	62
28	0.5317	0.5354	0.5392	0.5430	0.5467	0.5505	0.5543	4	8	11	15	19	61
29	0.5543	0.5581	0.5619	0.5658	0.5696	0.5735	0.5774	4	8	12	15	19	60
30	0.5774	0.5812	0.5851	0.5890	0.5930	0.5969	0.6009	4	8	12	16	20	59
31	0.6009	0.6048	0.6088	0.6128	0.6168	0.6208	0.6249	4	8	12	16	20	58
32	0.6249	0.6289	0.6330	0.6371	0.6412	0.6453	0.6494	4	8	12	16	20	57
33	0.6494	0.6536	0.6577	0.6619	0.6661	0.6703	0.6745	4	8	13	17	21	56
34	0.6745	0.6787	0.6830	0.6873	0.6916	0.6959	0.7002	4	9	13	17	21	55
35	0.7002	0.7046	0.7089	0.7133	0.7177	0.7221	0.7265	4	9	13	18	22	54
36	0.7265	0.7310	0.7355	0.7400	0.7445	0.7490	0.7536	5	9	14	18	23	53
37	0.7536	0.7581	0.7627	0.7673	0.7720	0.7766	0.7813	5	9	14	18	23	52
38	0.7813	0.7860	0.7907	0.7954	0.8002	0.8050	0.8098	5	10	14	19	24	51
39	0.8098	0.8146	0.8195	0.8243	0.8292	0.8342	0.8391	5	10	15	20	24	50
40	0.8391	0.8441	0.8491	0.8541	0.8591	0.8642	0.8693	5	10	15	20	25	49
41	0.8693	0.8744	0.8796	0.8847	0.8899	0.8952	0.9004	5	10	16	21	26	48
42	0.9004	0.9057	0.9110	0.9163	0.9217	0.9271	0.9325	5	11	16	21	27	47
43	0.9325	0.9380	0.9435	0.9490	0.9545	0.9601	0.9657	6	11	17	22	28	46
44	0.9657	0.9713	0.9770	0.9827	0.9884	0.9942	1.0000	6	11	17	23	29	45
	60°	50°	40°	30°	20°	10°	0°						Degrees

TANGENTS AND COTANGENTS

Deg-grees	*0°*	*10°*	*20°*	*30°*	*40°*	*50°*	*60°*	*1°*	*2°*	*3°*	*4°*	*5°*	
45	1.0000	0.0060	1.0117	1.0176	1.0237	1.0295	1.0355	6	12	18	25	30	44
46	1.0355	1.0417	1.0477	1.0538	1.0600	1.0661	1.0724	6	12	18	25	31	43
47	1.0724	1.0786	1.0850	1.0913	1.0976	1.1041	1.1106	6	13	19	25	32	42
48	1.1106	1.1171	1.1237	1.1303	1.1369	1.1436	1.1504	7	13	20	26	33	41
49	1.1504	1.1571	1.1640	1.1708	1.1778	1.1847	1.1918	7	14	21	27	34	40
50	1.1918	1.1989	1.2059	1.2131	1.2203	1.2275	1.2349	7	14	22	29	36	39
51	1.2349	1.2423	1.2497	1.2572	1.2647	1.2723	1.2799	8	15	23	30	38	38
52	1.2799	1.2877	1.2954	1.3032	1.3110	1.3191	1.3270	8	16	23	31	39	37
53	1.3270	1.3352	1.3432	1.3514	1.3597	1.3680	1.3764	8	16	25	33	41	36
54	1.3764	1.3848	1.3933	1.4019	1.4105	1.4193	1.4281	9	17	26	34	43	35
55	1.4281	1.4371	1.4460	1.4550	1.4641	1.4733	1.4826	9	18	27	36	45	34
56	1.4826	1.4920	1.5012	1.5108	1.5204	1.5301	1.5399	10	19	29	38	48	33
57	1.5399	1.5458	1.5597	1.5697	1.5797	1.5900	1.6003	10	20	30	40	50	32
58	1.6003	1.6107	1.6213	1.6319	1.6426	1.6533	1.6643	11	21	32	43	53	31
59	1.6643	1.6754	1.6865	1.6977	1.7090	1.7205	1.7311	11	23	34	45	56	30
60	1.7321	1.7439	1.7556	1.7675	1.7795	1.7917	1.8040	12	24	36	48	60	29
61	1.8040	1.8166	1.8291	1.8418	1.8546	1.8676	1.8807	13	26	38	51	64	28
62	1.8807	1.8942	1.9074	1.9210	1.9347	1.9485	1.9226	14	27	41	55	68	27
63	1.9626	1.9769	1.9912	2.0057	2.0203	2.0352	2.0503	15	29	44	58	73	26
64	2.0503	2.0657	2.0809	2.0965	2.1123	2.1282	2.1445	16	31	47	63	77	25
65	2.1445	2.1611	2.1776	2.1943	2.2113	2.2285	2.2460	17	34	51	68	85	24
66	2.2460	2.2640	2.2818	2.2998	2.3183	2.3369	2.3559	18	37	55	74	92	23
67	2.3559	2.3752	2.3946	2.4142	2.4343	2.4545	2.4751	20	40	60	79	99	22
68	2.4751	2.4963	2.5172	2.5386	2.5604	2.5825	2.6051	22	43	65	87	108	21
69	2.6051	2.6282	2.6511	2.6746	2.6984	2.7226	2.7475	24	47	71	95	118	20
70	2.7475	2.7729	2.7981	2.8239	2.8501	2.8768	2.9042	26	52	78	104	130	19
71	2.9042	2.9323	2.9602	2.9887	3.0176	3.0473	3.0777	29	58	87	115	144	18
72	3.0777	3.1080	3.1398	3.1716	3.2039	3.2369	3.2709	32	64	96	129	161	17
73	3.2709	3.3058	3.3404	3.3759	3.4121	3.4492	3.4874	36	72	108	144	180	16
74	3.4874	3.5267	3.5658	3.6059	3.6467	3.6888	3.7321	41	82	122	162	203	15
75	3.7321	3.7769	3.8219	3.8667	3.9133	3.9614	4.0108	46	94	139	186	232	14
76	4.0108	4.0622	4.1129	4.1653	4.2190	4.2742	4.3315	53	107	160	214	267	13
77	4.3315	4.3897	4.4497	4.5107	4.5731	4.6376	4.7046	62	124	186	248	310	12
78	4.7046	4.7745	4.8434	4.9152	4.9886	5.0650	5.1446	73	146	219	292	365	11
79	5.1446	5.2279	5.3099	5.3955	5.4836	5.5753	5.6713	87	175	262	350	437	10
80	5.671	5.769	5.871	5.976	6.084	6.197	6.314						9
81	6.314	6.435	6.561	6.691	6.827	6.968	7.115						8
82	7.115	7.269	7.429	7.596	7.770	7.953	8.144						7
83	8.144	8.345	8.556	8.777	9.010	9.255	9.514						6
84	9.514	9.788	10.078	10.385	10.712	11.059	11.430						5
85	11.430	11.826	12.251	12.706	13.197	13.727	14.301						4
86	14.301	14.924	15.605	16.305	17.169	18.075	19.081						3
87	19.081	20.206	21.470	22.904	24.542	26.432	28.636						2
88	28.636	31.242	34.368	38.188	42.964	49.104	57.290						1
89	57.290	68.750	85.940	114.59	171.89	343.77	∞						0
	60°	50°	40°	30°	20°	10°							Deg-grees